ERROR CONTROL CODING

ERROR CONTROL CODING
From Theory to Practice

Peter Sweeney
University of Surrey, Guildford, UK

JOHN WILEY & SONS, LTD

Other Wiley Editorial Offices

John Wiley & Sons, Inc., 605 Third Avenue, New York, NY 10158–0012, USA

Jossey-Bass, 989 Market Street, San Francisco, CA 94103–1741, USA

Wiley-VCH Verlag GmbH, Pappelallee 3, D-69469 Weinheim, Germany

John Wiley & Sons Australia, Ltd., 33 Park Road, Milton, Queensland 4064, Australia

John Wiley & Sons (Asia) Pte Ltd., 2 Clementi Loop #02–01, Jin Xing Distripark, Singapore 129809

John Wiley & Sons Canada, Ltd., 22 Worcester Road, Etobicoke, Ontario, Canada M9W 1L1

British Library Cataloguing in Publication Data

A catalogue record for this book is available from the British Library

ISBN 0 470 84356 X

Typeset in 10/12pt Times by Kolam Information Services Pvt. Ltd, Pondicherry, India
Printed and bound in Great Britain by TJ International, Padstow, Cornwall
This book is printed on acid-free paper responsibly manufactured from sustainable forestry
in which at least two trees are planted for each one used for paper production.

Contents

1

The principles of coding in digital communications

1.1 ERROR CONTROL SCHEMES

Error control coding is concerned with methods of delivering information from a source to a destination with a minimum of errors. As such it can be seen as a branch of information theory and traces its origins to Shannon's work in the late 1940s. The early theoretical work indicates what is possible and provides some insights into the general principles of error control. On the other hand, the problems involved in finding and implementing codes have meant that the practical effects of employing coding are often somewhat different from what was originally expected.

Shannon's work showed that any communication channel could be characterized by a capacity at which information could be reliably transmitted. At any rate of information transmission up to the channel capacity, it should be possible to transfer information at error rates that can be reduced to any desired level. Error control can be provided by introducing redundancy into transmissions. This means that more symbols are included in the message than are strictly needed just to convey the information, with the result that only certain patterns at the receiver correspond to valid transmissions. Once an adequate degree of error control has been introduced, the error rates can be made as low as required by extending the length of the code, thus averaging the effects of noise over a longer period.

Experience has shown that to find good long codes with feasible decoding schemes is more easily said than done. As a result, practical implementations may concentrate on the improvements that can be obtained, compared with uncoded communications. Thus the use of coding may increase the operational range of a communication system, reduce the error rates, reduce the transmitted power requirements or obtain a blend of all these benefits.

Apart from the many codes that are available, there are several general techniques for the control of errors, and the choice will depend on the nature of the data and the user's requirements for error-free reception. The most complex techniques fall into the category of forward error correction, where it is assumed that a code capable of correcting any errors will be used. Alternatives are to detect errors and request retransmission, which is known as retransmission error control, or to use inherent redundancy to process the erroneous data in a way that will make the errors subjectively important, a method known as error concealment.

This chapter first looks at the components of a digital communication system. Sections 1.3 to 1.8 then look in more detail at each of the components. Section 1.8 gives a simple example of a code that is used to show how error detection and correction may in principle be achieved. Section 1.9 discusses the performance of error correcting codes and Section 1.10 looks at the theoretical performance available. A number of more advanced topics are considered in Sections 1.11 to 1.14, namely coding for bandwidth-limited conditions, coding for burst errors, multistage coding (known as *concatenation*) and the alternatives to forward error correction. Finally, Section 1.15 summarizes the various considerations in choosing a coding scheme.

1.2   ELEMENTS OF DIGITAL COMMUNICATION SYSTEMS

A typical communication system incorporating coding is shown in Figure 1.1. Error control coding is applied after the source information is converted into digital format by the source encoder. The separation between coding and modulation is conventional, although it will be found later that there are instances where the two must be designed together. At the receiver, the operations are carried out in reverse order relative to the transmitter.

The functions are described in more detail in the following sections.

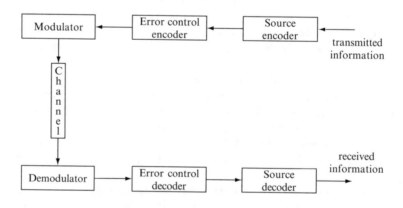

Figure 1.1   Coded communication system

1.3   SOURCE ENCODING

Information is given a digital representation, possibly in conjunction with techniques for removal of any inherent redundancy within the data. The amount of information contained in any message is defined in terms of the probability p that the message is selected for transmission. The information content H, measured in bits, is given by

$$H = \log_2 (1/p)$$

For example, a message with a 1% chance of being selected would contain approximately 6.64 bits of information.

If there are M messages available for selection and the probability of message m is denoted p_m, the average amount of information transferred in a message is

$$\overline{H} = \sum_{m=0}^{M-1} p_m \log_2 \left(\frac{1}{p_m}\right)$$

subject to the constraint that $\sum_{m=0}^{M-1} p_m = 1$.

If the messages are equiprobable, i.e. $p_m = 1/M$, then the average information transferred is just $\log_2 (M)$. This is the same as the number of bits needed to represent each of the messages in a fixed-length coding scheme. For example, with 256 messages an 8-bit code can be used to represent any messages and, if they are equally likely to be transmitted, the information content of any message is also 8 bits.

If the messages are not equally likely to be transmitted, then the average information content of a message will be less than $\log_2 (M)$ bits. It is then desirable to find a digital representation that uses fewer bits, preferably as close as possible to the average information content. This may be done by using variable length codes such as Huffman codes or arithmetic codes, where the length of the transmitted sequence matches as closely as possible the information content of the message. Alternatively, for subjective applications such as speech, images or video, lossy compression techniques can be used to produce either fixed-length formats or variable-length formats. The intention is to allow the receiver to reconstitute the transmitted information into something that will not exactly match the source information, but will differ from it in a way that is subjectively unimportant.

1.4   ERROR CONTROL CODING

Error control coding is in principle a collection of digital signal processing techniques aiming to average the effects of channel noise over several transmitted signals. The amount of noise suffered by a single transmitted symbol is much less predictable than that experienced over a longer interval of time, so the noise margins built into the code are proportionally smaller than those needed for uncoded symbols.

An important part of error control coding is the incorporation of redundancy into the transmitted sequences. The number of bits transmitted as a result of the error correcting code is therefore greater than that needed to represent the information. Without this, the code would not even allow us to detect the presence of errors and therefore would not have any error controlling properties. This means that, in theory, any incomplete compression carried out by a source encoder could be regarded as having error control capabilities. In practice, however, it will be better to compress the source information as completely as possible and then to re-introduce redundancy in a way that can be used to best effect by the error correcting decoder.

The encoder is represented in Figure 1.2. The information is formed into frames to be presented to the encoder, each frame consisting of a fixed number of symbols. In most cases the symbols at the input of the encoder are bits; in a very few cases symbols consisting of several bits are required by the encoder. The term symbol will be used to maintain generality.

To produce its output, the encoder uses the symbols in the input frame and possibly those in a number of previous frames. The output generally contains more symbols than the input, i.e. redundancy has been added. A commonly used descriptor of a code is the code rate (R) which is the ratio of input to output symbols in one frame. A low code rate indicates a high degree of redundancy, which is likely to provide more effective error control than a higher rate, at the expense of reducing the information throughput.

If the encoder uses only the current frame to produce its output, then the code is called a (n, k) block code, with the number of input symbols per frame designated k and the corresponding number of output symbols n. If the encoder remembers a number of previous frames and uses them in its algorithm, then the code is called a tree code and is usually a member of a subset known as convolutional codes. In this case the number of symbols in the input frame will be designated k_0 with n_0 symbols in the output frame. The encoder effectively performs a sliding window across the data moving in small increments that leave many of the same symbols still within the encoder window, as shown in Figure 1.3. The total length of the window, known as

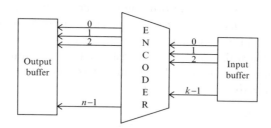

Figure 1.2   Encoder

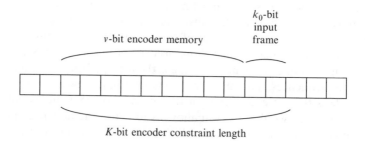

Figure 1.3   Sliding window for tree encoder

the *input constraint length* (K), consists of the input frame of k_0 symbols plus the number of symbols in the memory. This latter parameter is known as *memory constraint length* (v).

In more complex systems the encoding may consist of more than one stage and may incorporate both block and convolutional codes and, possibly, a technique known as interleaving. Such systems will be considered in later sections.

One property that will be shared by all the codes in this book is *linearity*. If we consider a linear system we normally think in terms of output being proportional to input (scaling property). For a linear system we can also identify on the output the sum of the separate components deriving from the sum of two different signals at the input (superposition property). More formally, if the system performs a function f on an input to produce its output, then

$$f(c\mathbf{x}) = c \times f(\mathbf{x}) \quad \text{(scaling)}$$

$$f(\mathbf{x} + \mathbf{y}) = f(\mathbf{x}) + f(\mathbf{y}) \quad \text{(superposition)}$$

where c is a scalar quantity, $\mathbf{x}$ and $\mathbf{y}$ are vectors.

Now the definition of a linear code is less restrictive than this, in that it does not consider the mapping from input to output, but merely the possible outputs from the encoder. In practice, however, a linear system will be used to generate the code and so the previous definition will apply in all real-life cases.

The standard definition of a linear code is as follows:

- Multiplying a code sequence by a valid scalar quantity produces a code sequence.

- Adding two code sequences produces a code sequence.

The general rules to be followed for multiplication and addition are covered in Chapter 5 but for binary codes, where the only valid scalars are 0 and 1, multiplication of a value by zero always produces zero and multiplication by 1 leaves the value unchanged. Addition is carried out as a modulo-2 operation, i.e. by an exclusive-OR function on the values.

A simple example of a linear code will be given in Section 1.8. Although the definition of a linear code is less restrictive than that of a linear system, in practice linear codes will always be produced by linear systems. Linear codes must contain the all-zero sequence, because multiplying any code sequence by zero will produce an all-zero result.

1.5 MODULATION

The modulator can be thought of as a kind of digital to analogue converter, preparing the digital code-stream for the real, analogue world. Initially the digital stream is put into a *baseband* representation, i.e. one in which the signal changes at a rate comparable with the rate of the digital symbols being represented. A convenient representation is the *Non Return to Zero* (NRZ) format, which represents bits by signal levels of $+V$ or $-V$ depending on the bit value. This is represented in Figure 1.4.

Figure 1.4   Binary NRZ stream

Although it would be possible to transmit this signal, it is usual to translate it into a higher frequency range. The reasons for this include the possibility of using different parts of the spectrum for different transmissions and the fact that higher frequencies have smaller wavelengths and need smaller antennas. For most of this text, it will be assumed that the modulation is produced by multiplying the NRZ baseband signal by a sinusoidal carrier whose frequency is chosen to be some multiple of the transmitted bit rate (so that a whole number of carrier cycles are contained in a single-bit interval). As a result, the signal transmitted over a single-bit interval is either the sinusoidal carrier or its inverse. This scheme is known as *Binary Phase Shift Keying* (BPSK).

It is possible to use a second carrier at 90° to the original one, modulate it and add the resulting signal to the first. In other words, if the BPSK signal is $\pm \cos(2\pi f_c t)$, where f_c is the carrier frequency and t represents time, the second signal is $\pm \sin(2\pi f_c t)$ and the resultant is

$$s(t) = \sqrt{2}\cos(2\pi f_c t + i\pi/4) \quad i = -3, -1, +1, +3$$

This is known as *Quadriphase Shift Keying* (QPSK) and has the advantage over BPSK that twice as many bits can be transmitted in the same time and the same bandwidth, with no loss of resistance to noise. The actual bandwidth occupied by the modulation depends on implementation details, but is commonly taken to be 1 Hz for one bit per second transmission rate using BPSK or 0.5 Hz using QPSK.

A phase diagram of QPSK is shown in Figure 1.5. The mapping of the bit values onto the phases assumes that each of the carriers is independently modulated using alternate bits from the coded data stream. It can be seen that adjacent points in the diagram differ by only one bit because the phase of only one of the two carriers has changed. A mapping that ensures that adjacent points differ by only one bit is known as *Gray Coding*.

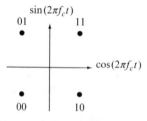

Figure 1.5   Gray-coded QPSK phase diagram

Other possible modulations include *Frequency Shift Keying* (FSK), in which the data determines the frequency of the transmitted signal. The advantage of FSK is simplicity of implementation, although the resistance to noise is less than BPSK or QPSK. There are also various modulations of a type known as *Continuous Phase Modulation*, which minimize phase discontinuities between transmitted waveforms to improve the spectral characteristics produced by nonlinear power devices.

In bandwidth-limited conditions, multilevel modulations may be used to achieve higher bit rates within the same bandwidth as BPSK or QPSK. In *M-ary Phase Shift Keying* (MPSK) a larger number of transmitted phases is possible. In *Quadrature Amplitude Modulation* (QAM) a pair of carriers in phase quadrature are each given different possible amplitudes before being added to produce a transmitted signal with different amplitudes as well as phases. QAM has more noise resistance than equivalent MPSK, but the variations in amplitude may cause problems for systems involving nonlinear power devices. Both QAM and MPSK require special approaches to coding which consider the code and the modulation together.

1.6 THE CHANNEL

The transmission medium introduces a number of effects such as attenuation, distortion, interference and noise, making it uncertain whether the information will be received correctly. Although it is easiest to think in terms of the channel as introducing errors, it should be realized that it is the effects of the channel on the demodulator that produce the errors.

The way in which the transmitted symbols are corrupted may be described using the following terms:

* Memoryless channel – the probability of error is independent from one symbol to the next.

* Symmetric channel – the probability of a transmitted symbol value i being received as a value j is the same as that of a transmitted symbol value j being received as i, for all values of i and j. A commonly encountered example is the binary symmetric channel (BSC) with a probability p of bit error, as illustrated in Figure 1.6.

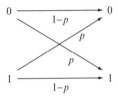

Figure 1.6 Binary symmetric channel

- Additive White Gaussian Noise (AWGN) channel – a memoryless channel in which the transmitted signal suffers the addition of wide-band noise whose amplitude is a normally (Gaussian) distributed random variable.

- Bursty channel – the errors are characterized by periods of relatively high symbol error rate separated by periods of relatively low, or zero, error rate.

- Compound (or diffuse) channel – the errors consist of a mixture of bursts and random errors. In reality all channels exhibit some form of compound behaviour.

Many codes work best if errors are random and so the transmitter and receiver may include additional elements, an *interleaver* before the modulator and a *deinterleaver* after the demodulator to randomize the effects of channel errors. This will be discussed in Section 1.12.

1.7 DEMODULATION

1.7.1 Coherent demodulation

The demodulator attempts to decide on the values of the symbols that were transmitted and pass those decisions on to the next stage. This is usually carried out by some sort of correlation with replicas of possible transmitted signals. Consider, for example, the case of BPSK. The correlation is with a single fixed phase of the carrier, producing either a positive or a negative output from the detector. In the absence of noise, the detected signal level can be taken as $\pm\sqrt{E_r}$ where E_r is the energy in each received bit. The effect of an AWGN channel will be to add a noise level n sampled from a Gaussian distribution of zero mean and standard deviation σ. The probability density is given by

$$p(n)\, dn = \frac{1}{\sigma\sqrt{2\pi}} e^{-n^2/2\sigma^s}$$

Gaussian noise has a flat spectrum and the noise level is often described by its *Single-sided Noise Power Spectral Density*, which is written N_0. The variance, σ^2, of the Gaussian noise, integrated over a single-bit interval, will be $N_0/2$. In fact it can be considered that there is a total noise variance of N_0 with half of this acting either in phase or in antiphase to the replica and the other half in phase quadrature, therefore not affecting the detector. The performance of Gray-coded QPSK is therefore exactly the same as BPSK because the two carriers can be demodulated as independent BPSK signals, each affected by independent Gaussian noise values with the same standard deviation.

The demodulator will make its decision based on the sign of the detected signal. If the received level is positive, it will assume that the value corresponding to the replica was transmitted. If the correlation was negative, it will assume that the other value was transmitted. An error will occur, therefore, if the noise-free level is $-\sqrt{E_r}$

and a noise value greater than $+\sqrt{E_r}$ is added, or if the noise-free level is $+\sqrt{E_r}$ and a noise value less than $-\sqrt{E_r}$ is added. Considering only the former case, we see that the probability of error is just the probability that the Gaussian noise has a value greater than $+\sqrt{E_r}$:

$$p = \frac{1}{\sqrt{\pi N_0}} \int_{\sqrt{E_r}}^{\infty} e^{-n^2/N_0} \, dn$$

Substituting $t = n/\sqrt{N_0}$ gives

$$p = \frac{1}{\sqrt{\pi}} \int_{\sqrt{\frac{E_b}{N_0}}}^{\infty} e^{-t^2/N_0} \, dt$$

$$p = \frac{1}{2} erfc \sqrt{\frac{E_b}{N_0}} \quad \text{where} \quad erfc(x) = \frac{2}{\sqrt{\pi}} \int_{t}^{\infty} e^{-t^2} \, dt \qquad (1.1)$$

The function $erfc(x)$ is known as the complementary error function and its values are widely available in tabulated form. Note that the maximum value of $erfc(x)$ is 1.0, so that the maximum bit error probability is 0.5. This makes sense because we could guess bit values with 50% probability without attempting to receive them at all.

1.7.2 Differential demodulation

One further complication is commonly encountered with BPSK or QPSK transmissions. In the absence of other information, it is impossible for the receiver to determine absolute phase so as to know which of the two phases represents the value 0 and which represents 1. This is because delay in the transmission path, which is equivalent to phase shift, will not be known. The representation of bit values is therefore often based on the difference between phases. Depending on the precise demodulation method, this is known either as *Differentially Encoded Phase Shift Keying* (DEPSK) or as *Differential Phase Shift Keying* (DPSK). The two are identical from the modulation point of view, with the bit value 0 normally resulting in a change of phase and the bit value 1 resulting in the phase remaining the same. The receiver may not know absolute phase values, but should be able to tell whether the phase has changed or remained the same. The differences in the demodulator implementation may be summarized as follows:

- *DEPSK* – The demodulator maintains a replica of one of the two carrier phases and correlates the received signal with this replica as for normal PSK. It then compares the sign of the correlation with the previous correlation value; a change of sign indicates data bit 0 and the same sign indicates data bit 1. Compared with PSK, there will now be a bit error either when the phase is received wrongly and the previous phase was correct or when the phase is received correctly and the previous phase was wrong. Thus noise that would cause a single-bit error in a BPSK demodulator will cause two consecutive bit errors in the DEPSK demodulator and the bit error probability is approximately twice the above BPSK expression.

- *DPSK* – The demodulator uses the previously received phase as the replica for the next bit. Positive correlation indicates data value 1, negative correlation indicates data value 0. The bit errors again tend to correlate in pairs, but the overall performance is worse. In fact the bit error probability of DPSK follows a different shape of curve:

$$p = \frac{1}{2} e^{-E_r/N_0}$$

1.7.3   Soft-decision demodulation

In some cases the demodulator's decision will be easy; in other cases it will be difficult. In principle if errors are to be corrected it is better for the demodulator to pass on the information about the certainty of its decisions because this might assist the decoder in pinpointing the positions of the likely errors; this is called soft-decision demodulation. We could think of it as passing on the actual detected level as a real number, although in practice it is likely to have some sort of quantization. Eight-level quantization is found to represent a good compromise between complexity and performance.

Since the purpose of soft decisions is to assist decoding, it is useful to relate the demodulator output to probabilistic measures of performance. One commonly adopted measure is known as the *log-likelihood ratio*, defined as $\log[p(1|r_i)/p(0|r_i)]$. This metric is required in an important decoding method to be described in Chapter 10 and can be used for other decoding methods too. The computation of the value may appear difficult, however we note that

$$\log\left[\frac{p(1|r_i)}{p(0|r_i)}\right] = \log\left[\frac{p(r_i|1)}{p(r_i|0)}\right] + \log\left[\frac{p(1)}{p(0)}\right]$$

Assuming that values 0 and 1 are equiprobable, $\log[p(1)/p(0)] = 0$ and so the assigned bit value for received level r_i is equal to $\log[p(r_i|1)/p(r_i|0)]$. This value can be calculated given knowledge of the signal level and the noise statistics. Note that it ranges from $-\infty$ (certain 0) to $+\infty$ (certain 1).

Assuming that we have Gaussian noise, the probability density function at a received value r_i from a noise-free received value x is

$$p(r_i) = \frac{1}{\sqrt{\pi N_0}} e^{-(r_i - x)^2/N_0}$$

The appropriate values of x for bit values 1 and 0 are $+\sqrt{E_r}$ and $-\sqrt{E_r}$. Thus the log-likelihood ratio is proportional to $\log[e^{-(r_i - \sqrt{E_r})^2/N_0}/e^{-(r_i + \sqrt{E_r})^2/N_0}]$.

Now

$$\log\left[\frac{e^{-(r_i-\sqrt{E_r})^2/N_0}}{e^{-(r_i+\sqrt{E_r})^2/N_0}}\right] = -\frac{(r_i-\sqrt{E_r})^2}{N_0} + \frac{(r_i+\sqrt{E_r})^2}{N_0}$$

Hence we find that

$$\log\left[\frac{p(r_i|1)}{p(r_i|0)}\right] = \frac{4r_i\sqrt{E_r}}{N_0} = \frac{4r_i}{\sqrt{E_r}}\frac{E_r}{N_r}$$

In other words, the log-likelihood ratio is linear with the detected signal level and is equal to the channel E_r/N_0, multiplied by four times the detected signal (normalized to make the noise-free levels equal to $+/-1$).

Note that the mapping adopted here from code bit values to detected demodulator levels is opposite to that conventionally used in other texts. The conventional mapping is that bit value 0 maps onto $+1$ and bit value 1 onto -1. The advantage is that the exclusive-OR operation in the digital domain maps onto multiplication in the analog domain. The disadvantage is the potential confusion between bit value 1 and analog value $+1$.

Because of the linearity of the log-likelihood ratio, the quantization boundaries of the demodulator can be set in roughly linear steps. The question remains, however, as to what size those steps should be. It can be shown that, for Q-level quantization, the optimum solution is one that minimizes the value of

$$\sum_{j=0}^{Q-1}\sqrt{p(j|1)p(j|0)}$$

where $p(j|c)$ represents the probability of a received value j given that symbol c was transmitted. Massey[1] described an iterative method of finding the optimum solution with nonuniform arrangement of boundaries, but the above value can easily be calculated for different linear spacings to find an approximate optimum. For example, with E_r/N_0 around 2 dB, it is found that uniformly spaced quantization boundaries are close to optimum if the spacing is $1/3$, i.e. the boundaries are placed at $-1, -2/3$ $-1/3, 0, +1/3, +2/3, +1$. The use of such a scheme will be described in Section 1.8.2.

1.8   DECODING

The job of the decoder is to decide what the transmitted information was. It has the possibility of doing this because only certain transmitted sequences, known as code-words, are possible and any errors are likely to result in reception of a non-code sequence. On a memoryless channel, the best strategy for the decoder is to compare the received sequence with all the codewords, taking into account the confidence in the received symbols, and select the codeword which is closest to the received sequence as discussed above. This is known as *maximum likelihood decoding*.

1.8.1   Encoding and decoding example

Consider, for example, the block code shown in Table 1.1. This code is said to be
systematic, meaning that the codeword contains the information bits and some other
bits known as *parity checks* because they have been calculated in some way from the
information. It can be seen that any codeword differs in at least three places from any
other codeword. This value is called the *minimum Hamming distance* or, more briefly,
minimum distance of the code. Consequently, if a single bit is wrong, the received
sequence is still closer to the transmitted codeword, but if two or more bits are
wrong, then the received sequence may be closer to one of the other codewords.

   This code is linear and for any linear code it is found that the *distance structure* is
the same from any codeword. For this example, starting from any codeword there are
two sequences at a distance of 3 and one at a distance of 4. Thus the code properties
and the error-correction properties are independent of the sequence transmitted. As a
consequence, the minimum distance of the code can be found by comparing each
nonzero sequence with the all-zero sequence, finding the nonzero codeword with the
smallest number nonzero symbols. The count of nonzero symbols is known as the
weight of the sequence and the minimum weight of the code is equal to the minimum
distance.

   Let us now assume that information 10 has been selected and that the sequence
10101 is therefore transmitted. Let us also assume that the received bits are hard-
decision quantized. If the sequence is received without error, it is easy to identify it in
the table and to decode. If there are errors, however, things will be more difficult and
we need to measure the number of differences between the received sequence and
each codeword. The measure of difference between sequences is known as *Hamming
distance*, or simply as distance between the sequences. Consider first the received
sequence 00101. The distance to each codeword is shown in Table 1.2.

   In this case we can see that we have a clear winner. The transmitted sequence has
been selected as the most likely and the decoding is correct.

Table 1.1   Example block code

Information	Codeword
00	00000
01	01011
10	10101
11	11110

Table 1.2   Distances for sequence 00101

Codeword	Distance
00000	2
01011	3
10101	1
11110	4

The previous example had an error in an information bit, but the result will be the same if a parity check bit is wrong. Consider the received sequence 10111. The distances are shown in Table 1.3. Again the sequence 10101 is chosen. Further examples are left to the reader, but it will be found that any single-bit error can be recovered, regardless of the position or the codeword transmitted.

Now let us consider what happens if there are two errors. It will be found that there are two possibilities.

Firstly, consider the received sequence 11111. The distances are shown in Table 1.4. In this case, the codeword 11110 is chosen, which is wrong. Moreover, the decoder has decided that the final bit was wrong when in fact it was correct. Because there are at least three differences between any pair of codewords, the decoder has made an extra error on top of the two made by the channel, in effect making things worse.

Finally, consider the received sequence 11001, whose distances to the codewords are shown in Table 1.5. In this case, there are two problems in reaching a decision. The first, and obvious, problem is that there is no clear winner and, in the absence of other information, it would be necessary to choose randomly between the two most likely codewords. Secondly, we predicted at the outset that only single errors would be correctable and the decoder may have been designed in such a way that it refuses to decode if there is no codeword within a distance 1 of the received sequence. The likely outcome for this example, therefore, is that the decoder will be unable to

Table 1.3   Distances for sequence 10111

Codeword	Distance
00000	4
01011	3
10101	1
11110	2

Table 1.4   Distances for sequence 11111

Codeword	Distance
00000	5
01011	2
10101	2
11110	1

Table 1.5   Distances for sequence 11001

Codeword	Distance
00000	3
01011	2
10101	2
11110	3

choose the most likely transmitted codeword and will indicate to the user the presence of *detected uncorrectable errors*. This is an important outcome that may occur frequently with block codes.

1.8.2 Soft-decision decoding

The probability that a sequence $\mathbf{c}$ of length n was transmitted, given the received sequence $\mathbf{r}$, is $\prod_{i=0}^{n-1} p(c_i|r_i)$. We wish to maximize this value over all possible code sequences. Alternatively, and more conveniently, we take logarithms and find the maximum of $\sum_{i=0}^{n-1} \log[p(c_i|r_i)]$. This can be carried out by a correlation process, which is a symbol-by-symbol multiplication and accumulation, regarding the code bits as having values $+1$ or -1. Therefore we would be multiplying the assigned probability by 1 for a code bit of 1 and by -1 for a code bit of 0. For hard decisions, a codeword of length n at a distance d from the received sequence would agree in $n - d$ places and disagree in d places with the received sequence, giving a correlation metric of $2n - d$. Obviously choosing the codeword to maximize this metric would yield the same decoding result as the minimum distance approach.

Even with soft decisions, we can adopt a minimum distance view of decoding and minimize $\sum_{i=0}^{n-1} \{1 - \log[p(c_i|r_i)]\}$. The correlation and minimum distance approaches are again identical provided we have an appropriate measure of distance. If the received bits are given values v_i equal to $\log[p(1|r_i)]$, then the distance to a bit value 1 is $1 - v_i$, the distance to a bit value 0 is v_i and we maximize probability by minimizing this measure of distance over all codewords.

The maximization of probability can also be achieved by maximizing some other function that increases monotonically with it. This is the case for the log-likelihood ratio $\log[p(1|r_i)/p(0|r_i)]$. To decode, we can maximize $\sum_i c_i \log[p(r_i|1)/p(r_i|1)]$ where c_i is taken as having values ± 1. This again corresponds to carrying out a correlation of received log-likelihood ratios with code sequences.

As discussed in Section 1.7.3, it is likely that the received levels will be quantized. For 8-level quantization, it might be convenient to use some uniform set of metric values depending on the range within which the detected bit falls. Such a scheme is shown in Figure 1.7.

Bearing in mind the fact that the log-likelihood ratio is linear with the analog detected level from the demodulator, then the only deviation from an ideal 8-level

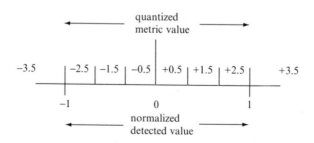

Figure 1.7 Quantization boundaries for soft-decision demodulation

quantization is that the end categories (000 and 111) extend to $-\infty$ and $+\infty$ and therefore should have larger metrics associated with them. The effect on performance, however, is negligible. For $E_r/N_0 = 2\,\text{dB}$, the optimum soft-decision metric values associated with this quantization arrangement are -3.85, -2.5, -1.5, -0.5, $+0.5$, $+1.5$, $+2.5$, $+3.85$. Therefore the proposed metrics of -3.5 to $+3.5$ are very close to optimum.

The assigned values can be scaled and offset in any convenient manner, so the scheme in Figure 1.7 is equivalent to having bit values of $(-7, -5, -3, -1, +1, +3, +5, +7)$ or $(0, 1, 2, 3, 4, 5, 6, 7)$. This last form is convenient for implementation of a 3-bit interface to the decoder.

Applying the correlation approach to a soft-decision case, the example in Table 1.4 might become a received sequence $+2.5\ +0.5\ +1.5\ +0.5\ +3.5$ with correlation values as shown in Table 1.6.

The maximum correlation value indicates the decoder decision. In this case, the decoder selects the correct codeword, illustrating the value of soft decisions from the demodulator.

Table 1.6   Correlations for soft-decision sequence $+2.5\ +0.5\ +1.5\ +0.5\ +3.5$

Codeword	Correlation
$-1\ -1\ -1\ -1\ -1$	-8.5
$-1\ +1\ -1\ +1\ +1$	$+0.5$
$+1\ -1\ +1\ -1\ +1$	$+6.5$
$+1\ +1\ +1\ +1\ -1$	$+1.5$

1.8.3   Alternative decoding approaches

Although conceptually very simple, the method described above is very complex to implement for many realistic codes where there may be very many codewords. As a result, other decoding methods will need to be studied. For example, the parity checks for the above code were produced according to very simple rules. Numbering the bits from left to right as bits 4 down to 0, bits 4 and 3 constitute the information and the parity bits are

$$\text{bit } 2 = \text{bit } 4$$
$$\text{bit } 1 = \text{bit } 3$$
$$\text{bit } 0 = \text{bit } 4 \oplus \text{bit } 3$$

The symbol $\oplus$ denotes modulo-2 addition or exclusive-OR operation. Considering only hard decisions, when a sequence is received, we can simply check whether the parity rules are satisfied and we can easily work out the implications of different error patterns. If there are no errors, all the parity checks will be correct. If there is a single-bit error affecting one of the parity bits, only that parity check will fail. If bit 4 is in error, parity bits 2 and 0 will be wrong. If bit 3 is in error, parity bits 1 and 0 will be wrong. If both parity bits 2 and 1 fail, the error is uncorrectable, regardless of whether parity bit 0 passes or fails.

We can now construct some digital logic to check the parity bits and apply the above rules to correct any correctable errors. It will be seen that applying the rules will lead to the same decodings as before for the examples shown. In the final example case, where the sequence 11001 was received, all three parity checks fail.

This type of decoding procedure resembles the methods applied for error correction to many block codes. Note, however, that it is not obvious how such methods can incorporate soft decisions from the demodulator. Convolutional codes, however, are decoded in a way that is essentially the same as the maximum likelihood method and soft decisions can be used.

1.9 CODE PERFORMANCE AND CODING GAIN

We saw earlier that we can obtain a theoretical expression for the bit error probability of BPSK or QPSK on the AWGN channel in terms of the ratio of energy per received bit to single-sided noise power spectral density, E_r/N_0. It is convenient to do the same for systems that employ coding, however we first have to solve a problem of comparability. Coding introduces extra bits and therefore we have to increase either the time to send a given message or else the bandwidth (by transmitting faster). Either case will increase the total noise in the message; in the first case because we get noise from the channel for a longer time, in the second case because more noise falls within the bandwidth.

The answer to this problem is to assess the error performance of the link in terms of E_b/N_0, the ratio of energy per bit of information to noise power spectral density. Thus when coding is added, the number of bits of information is less than the number of transmitted bits, resulting in an increase in E_b/N_0 relative to E_r/N_0. For example, if 100 bits of information are to be sent using a rate 1/2 code, 200 bits must be transmitted. Assuming that we maintain the transmitted bit rate and power, the energy in the message is doubled, but the amount of information remains the same. Energy per bit of information is therefore doubled, an increase of 3 dB. This increase acts as a penalty that the code must overcome if it is to provide real gains. The performance curve is built up in three stages as explained below.

As the first stage, the curve of bit error rate (BER) against E_b/N_0 (the same as E_r/N_0 in this case) is plotted for the modulation used. The value of E_b/N_0 is usually measured in dB and the bit error rate is plotted as a logarithmic scale, normally covering several decades, e.g. from 10^{-1} to 10^{-6}. The second stage is the addition of coding without consideration of the changes to bit error rates. For a fixed number of transmitted bits, the number of information bits is reduced, thus increasing the value of E_b/N_0 relative to E_r/N_0 by a factor $1/R$, or by $10 \log_{10}(1/R)$ dB. The third stage is to consider the effect of coding on bit error rates; this may be obtained either by simulation or by calculation. For every point on the uncoded performance curve, there will therefore be a corresponding point a fixed distance to the right of it on the coded performance curve showing a different, in many cases lower, bit error rate.

An example is shown in Figure 1.8 which shows the theoretical performance of a BPSK (or QPSK) channel, uncoded and with a popular rate 1/2 convolutional code. The code performance is plotted both with hard-decision demodulation and with unquantized soft decisions, i.e. real number output of detected level from the demodulator.

It can be seen that without coding, the value of E_b/N_0 needed to achieve a bit error rate of 10^{-5} is around 9.6 dB. This error rate can be achieved with coding at E_b/N_0 around 7.1 dB using hard-decision demodulation or around 4.2 dB using unquantized soft-decision demodulation. This is expressed by saying that the *coding gain* at a BER of 10^{-5} is 2.5 dB (hard-decision) or 5.4 dB (soft-decision). Real life decoding gains would not be quite so large. The use of 8-level, or 3-bit, quantization of the soft decisions reduces the gain by around 0.25 dB. There may also be other implementation issues that affect performance. Nevertheless, gains of 4.5 to 5 dB can be expected with this code.

The quoted coding gain must be attached to a desired bit error rate, which in turn will depend on the application. Note that good coding gains are available only for relatively low required bit error rates and that at higher error rates the gain may be negative (i.e. a loss). Note also that the quoted bit error rate is the error rate coming out of the decoder, *not* the error rate coming out of the demodulator. In the soft-decision example, the demodulator is working at E_r/N_0 around 1.2 dB, producing a BER of around 5×10^{-2} out of the demodulator.

If we know the minimum distance of a block code, or the value of an equivalent parameter called *free distance* for a convolutional code, we can find the *asymptotic*

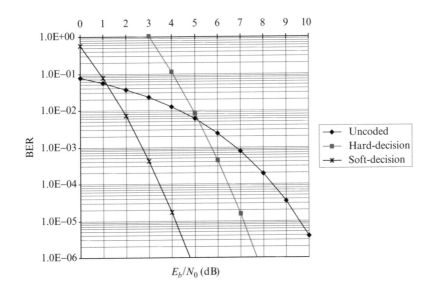

Figure 1.8 Performance of rate 1/2 convolutional code

coding gain, i.e. the gain that would be delivered if vanishingly small decoded error rates were required. For unquantized soft-decision decoding of a rate R code with distance d between the closest code sequences, the asymptotic gain is

$$G_{\text{asymptotic}} = 10 \log_{10} (Rd) \qquad (1.2)$$

If we have only hard decisions from the demodulator and can correct up to t errors then

$$G_{\text{asymptotic}} = 10 \log_{10} [R(t + 1)] \qquad (1.3)$$

From the earlier simple block code example, we can see that the expected value of t would be the largest value that is less than half the value of d. Thus the value of d in Equation (1.2) is just less than twice the value of $t + 1$ in Equation (1.3). Asymptotic coding gains are therefore almost 3 dB higher when unquantized soft-decision decoding is used. As stated above, the use of 8-level quantization reduces the gain by about 0.25 dB.

Although we have solved one issue of comparability by the use of E_b/N_0, there is another that is regularly ignored. If we look at an uncoded channel and a coded channel with the same BER, the characteristics will be completely different. On the AWGN channel, the errors will occur at random intervals. On the coded channel there will be extended error-free intervals interspersed with relatively dense bursts of errors when the decoder fails. Thus if we are interested in error rates on larger units of transmission, frames, packets or messages, the coded channel at the same BER will give fewer failures but more bit errors in corrupted sections of transmission. Assessing coding gain by comparing coded and uncoded channels with the same BER may therefore be unfair to the coded channel. For example, out of 100 messages sent, an uncoded channel might result in 10 message errors with one bit wrong in each. A coded channel might produce only one message error but 10 bit errors within that message. The bit error rates are the same, but the message error rate is better on the coded channel. Add to this the fact that the detection of uncorrectable errors is rarely taken into account in a satisfactory way (a significant issue for many block codes), coding regularly delivers benefits that exceed the theoretical figures.

1.10   INFORMATION THEORY LIMITS TO CODE PERFORMANCE

We have now seen the sort of benefits that coding provides in present day practice and the ways to find asymptotic coding gain based on knowledge of simple code parameters. As yet we have not seen how to do detailed error rate calculations as these require a more detailed knowledge of code structure. Nevertheless, it is worth making a comparison with the results obtained from Shannon's work on information theory to show that, in some respects, coded systems have still some way to go.

Shannon showed that, using an average of all possible codes of length n, the error rate over the channel is characterized by a probability of message error

$$P_e \leq e^{-nE(R_1)} \tag{1.4}$$

where E, which is a function of the information rate, is called the random coding error exponent. Any specific code will have its own error exponent and the greater the error exponent the better the code, but there are calculable upper and lower bounds to the achievable value of E. In particular, a positive error exponent is achievable provided R_I is less than some calculable value called the channel capacity. Provided a positive error exponent can be obtained, the way to achieve lower error probabilities is to increase the length of the code.

As was seen in Section 1.9, codes have a calculable asymptotic coding gain and thus at high signal-to-noise values the error rates reduce exponentially with E_b/N_0, as in the uncoded case. The error exponent is therefore proportional to E_b/N_0. The difficulty with known codes is maintaining the error exponent while the length is increased. All known codes produced by a single stage of encoding can hold their value of error exponent only by reducing the rate to zero as the code length increases towards infinity. For example, an orthogonal signal set, which can be achieved by Frequency Shift Keying or by means of a block code, is sometimes quoted as approaching the theoretical capacity on an AWGN channel as the signal set is expanded to infinity. Unfortunately the bandwidth efficiency or the code rate reduces exponentially at the same time. This limitation can be overcome by the use of multistage encoding, known as concatenation, although even then the error exponents are less than the theoretically attainable value. Nevertheless, concatenation represents the closest practicable approach to the predictions of information theory, and as such is a technique of increasing importance. It is treated in more detail in Chapters 9 and 10.

As the most widely available performance figures for error correcting codes are for the additive white Gaussian noise (AWGN) channel, it is interesting to look at the theoretical capacity of such a channel. The channel rate is given by the Shannon–Hartley theorem:

$$C = B \log_2 \left(1 + \frac{S}{N}\right) \tag{1.5}$$

where B is bandwidth, S is signal power and N is noise power within the bandwidth. This result behaves roughly as one might expect, the channel capacity increasing with increased bandwidth and signal-to-noise ratio. It is interesting to note, however, that in the absence of noise the channel capacity is not bandwidth-limited. Any two signals of finite duration are bound to show differences falling within the system bandwidth, and in the absence of noise those differences will be detectable.

Let $N = B \cdot N_0$ and $S = R_I E_b$ (N_0 is the single-sided noise power spectral density, R_I is rate of information transmission ($\leq C$) and E_b is energy per bit of information), then

$$C = B \log_2\left(1 + \frac{R_I E_b}{B N_0}\right)$$

In the limit of infinite bandwidth, using the fact that $\log_2(x) = \log_e(x)/\log_e 2$ gives

$$C = 1.44 B \log_e\left(1 + \frac{R_I E_b}{B N_0}\right)$$

As bandwidth approaches infinity, the channel capacity is given by

$$C \approx 1.44 R_I \frac{E_b}{N_0}$$

For transmission at the channel capacity, $(R_I = C)$:

$$\frac{E_b}{N_0} = \frac{1}{1.44} = -1.6\,\text{dB} \tag{1.6}$$

This means that we should be able to achieve reliable communications at the channel capacity with values of E_b/N_0 as low as -1.6 dB. The channel capacity is however proportional to the information rate; increasing the rate for a fixed value of E_b/N_0 increases the signal power and therefore the channel capacity. Thus at -1.6 dB we should be able to achieve reliable communications at any rate over an AWGN channel, provided we are willing to accept infinite bandwidth.

If instead we constrain the bandwidth and set $R_I = \eta B$, where η is bandwidth efficiency of the modulation/coding scheme, then

$$C = \frac{R_I}{\eta} \log_2\left(1 + \frac{\eta E_b}{N_0}\right)$$

For transmission at the channel capacity $(R_I = C)$, therefore

$$\frac{E_b}{N_0} = \frac{1}{\eta}(2^\eta - 1) \tag{1.7}$$

This value can be thought of as imposing an upper limit to the coding gain achievable by a particular coding and modulation scheme. The value of E_b/N_0 to deliver the desired error rate on the uncoded channel can be determined from the modulation performance, and the corresponding coded value must be at least that given by equation (1.7). In practice, these coding gains are difficult to achieve.

If we were to use a rate $1/2$ code on a QPSK channel, a fairly common arrangement, the value of η is around 1.0, giving $E_b/N_0 = 1\,(= 0\,\text{dB})$. As has been seen earlier, a rate $1/2$ convolutional code may need over 4.5 dB to deliver a BER of 10^{-5}. It therefore falls well short of the theoretical maximum gain.

It must be stressed that Shannon merely proved that it was possible by coding to obtain reliable communications at this rate. There is no benefit, however, in having a good code if one does not know how to decode it. Practical codes are designed with a feasible decoding method in mind and the problem of constructing long codes that can be decoded is particularly severe. This seems to be the main reason why approaching the Shannon performance has proved to be so difficult.

1.11 CODING FOR MULTILEVEL MODULATIONS

The standard modulation for satellite communications is QPSK, but 8-PSK or 16-PSK could be used to obtain 3 or 4 transmitted bits (respectively) per transmitted symbol. Unfortunately, this results in reduced noise immunity. With m bits per transmitted symbol, assuming that the energy per transmitted bit is maintained, the energy per transmitted symbol can increase by a factor of m relative to binary PSK. The distance between closest points in the constellation will, however, be proportional to $\sin(\pi/M)$, where $M = 2^m$, as shown in Figure 1.9, and the noise energy required to cause an error will depend on the square of this. The uncoded performance relative to binary PSK is therefore

$$G_m\,(\text{dB}) = 10\,\log_{10}\left[m\sin^2\left(\pi/2^m\right)\right]$$

The values are shown in Table 1.7.

As can be seen, there are severe losses associated with higher level constellations, making coding all the more important. The codes, however, need to be designed specifically for the constellation to maximize the distance in signal space, the *Euclidean Distance*, between code sequences.

Figure 1.9 Distance between MPSK constellation points

Table 1.7 Performance of uncoded MPSK

m	M	G_m (dB)
1	2	0.0
2	4	0.0
3	8	−3.6
4	16	−8.2
5	32	−13.2
6	64	−18.4
7	128	−23.8
8	256	−29.2

The principal approach to designing codes for this type of system is to take a constellation with m bits per symbol and to use a rate $(m - 1)/m$ code so that the information throughput will be the same as the uncoded constellation with $m - 1$ bits per symbol and the performances can be compared directly. Convolutional codes of this type are known as Ungerboeck codes and will be described in Chapter 2.

1.12 CODING FOR BURST-ERROR CHANNELS

Coding performance curves are regularly shown for the AWGN channel. There are two reasons why this is so. Firstly, burst-error mechanisms are often badly understood and there may be no generally accepted models that fit the real behaviour. The increasing importance of mobile communications where the channel does not remotely fit the AWGN model has, however, led to considerable advances in the modelling of non-Gaussian channels. The other reason is that most codes in use are primarily designed for random error channels. The only important codes where this is not the case are Reed Solomon codes which are constructed with multibit symbols and correct a certain number of symbol errors in each codeword. A burst of errors affecting several bits close together may affect only a few symbols of the code and be correctable, as shown in Figure 1.10. The symbols each consist of 4 bits and a burst spanning 8 bits containing 5 errors has affected only 3 symbols.

For the most part, we shall be faced with trying to make a random bit-error-correcting code work on a burst-error channel, and the technique that is used is interleaving. Essentially, this consists of reordering the bits before transmission (interleaving) and putting them back into the original order on reception (deinterleaving). As the error locations are affected only by the deinterleaving, they become scattered through the code-stream so that they appear as random errors to the decoder.

There are two main types of interleaving to consider, block interleaving and convolutional interleaving. Both will be explained as if they are being used with a block code, although both can be used with convolutional codes too.

Block interleaving is illustrated in Figure 1.11. Codewords are written into the columns of an array and the total number of columns, λ, is termed the *interleaving degree*. If a burst of errors spans no more than λ symbols, then there will be at most one error in each codeword. A code that can correct up to t errors could correct, for example, up to t bursts of length λ, one burst of length $(\lambda t$ or a mixture of shorter bursts and random errors.

Convolutional interleaving is shown in Figure 1.12. The codewords in the columns of the array are shifted through delays which differ for each symbol. Usually these are increasing by one for each row of the array. The order of symbols on the channel

Error location

Figure 1.10 Binary burst error on multibit symbols

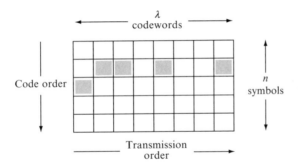

Figure 1.11   Block interleaving

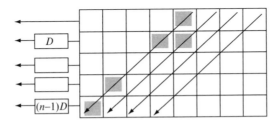

Figure 1.12   Convolutional interleaving

follows the diagonal sequence shown. Any burst of errors will affect symbols in the transmission stream as shown and it can be seen that the burst must exceed $n + 1$ symbols in length before it affects two symbols of the same codeword. If the delays are increasing by D for each symbol, then the separation of two symbols from the same codeword is $Dn + 1$. In effect this is the interleaving degree.

The main differences between the two types of interleaving are that the convolutional interleaver will extend the symbol stream through the presence of null values in the delay registers, but block interleaving will have more delay because of the need to fill the array before transmission can commence.

One might think that the block interleaver would introduce a delay of λn symbols, however it is possible to start transmission a little before the array is filled. The encoder must have $(\lambda - 1)n + 1$ symbols prepared by the time that λ symbols are transmitted; otherwise, the rightmost symbol of the top row will not be ready in time for transmission (assuming that symbol is transmitted the instant it is prepared). The delay is therefore $(\lambda - 1)n + 1 - \lambda = (\lambda - 1)(n - 1)$ symbols. The same delay will occur in the deinterleaver which writes the symbols into rows and decodes by column, giving an overall delay of $2(\lambda - 1)(n - 1)$ symbols.

The convolutional interleaver introduces $D + 2D + \cdots + (n - 1)D = n(n - 1)D/2$ dummy symbols into the stream. The deinterleaver applies a delay of $(n - 1)D$ to the top row, $(n - 2)D$ to the second row, etc., introducing the same number of dummy symbols. The overall delay is therefore $n(n - 1)D$. As the interleaving degree is $nD + 1$, the overall delay is $(\lambda - 1)(n - 1)$, half the value of the block interleaving.

1.13   MULTISTAGE CODING

The aim of making an effective long code is sometimes approached by multistage coding in which the overall code is constructed from simple components, thus providing a feasible approach to decoding. Examples of this type of approach include serial concatenation, in which information is first encoded by one code, the *outer code*, and then the encoded sequence is further encoded by a second code, the *inner code*. Reed Solomon codes are often used as outer codes because of their ability to correct the burst errors from the inner decoder. Another approach is the *product code* in which information is written into an array and the rows and columns are separately encoded.

In recent years other types of concatenation have become of interest in conjunction with iterative decoding techniques, where decoding of the second code is followed by one or more further decodings of both codes. In particular, iterative decoding is applied to *parallel concatenated codes*, namely the application of two systematic codes to a single-information stream to derive two independent sets of parity checks. This is the principle of the so-called *turbo codes* and other similar constructions, which are treated in Chapter 10.

1.14   ERROR DETECTION BASED METHODS

So far we have assumed that the intention is to correct all errors if possible; this is known as *Forward Error Correction* (FEC). We have, however, seen that detected uncorrectable errors are possible. In fact there may be good reasons not to attempt error correction provided we have some other way of dealing with erroneous data. Not attempting error correction will not make the maximum use of the received sequence, but it makes it less likely that there will be undetected errors and reduces the complexity at the receiver.

There are two main possibilities if errors are not to be corrected. The first approach is to use a reverse channel (where available) to call for retransmission. This is known as *Retransmission Error Control* (REC) or *Automatic Retransmission reQuest* (ARQ). The second approach is to process the data in such a way that the effect of errors is minimized. This is called *Error Concealment*.

1.14.1   ARQ strategies

The transmitter breaks the data into frames, each of which contains a block code used for error detection. The receiver sends back acknowledgements of correct frames and whenever it detects that a frame is in error it calls for retransmission. Often the transmitter will have carried on sending subsequent frames, so by the time it receives the call for retransmission (or fails to obtain an acknowledgement within a predefined interval) it will already have transmitted several more frames. It can then either repeat just the erroneous frame (*Selective Repeat ARQ*) or else go back to the point in the sequence where the frame error occurred and repeat all frames from that point regardless (*Go Back N ARQ*).

If Selective Repeat (SR-ARQ) is employed, the receiver must take responsibility for the correct ordering of the frames. It must therefore have sufficient buffering to reinsert the repeated frame into the stream at the correct point. Unfortunately, it is not possible to be sure how many repeats will be needed before the frame will be received. The protocols therefore need to be designed in such a way that the transmitter recognizes when the receiver's buffer is full and repeats not only erroneous frames but also those which will have been lost through buffer overflow.

Neglecting effects of finite buffers, assuming independence of errors from frame to frame and a frame error rate of p_f, the efficiency of SR-ARQ is

$$\eta_{SR} = (1 - p_f)\left(\frac{k}{n}\right) \tag{1.8}$$

where n is the total frame length and k is the amount of information in the frame. The difference between n and k in this case will not be purely the parity checks of the code. It will include headers, frame numbers and other fields required by the protocol.

For Go Back N (GBN-ARQ), there is no need for receiver buffering, but the efficiency is lower. Every time x frames are received correctly, followed by one in error, the transmitter goes on to frame $x + N$ before picking up the sequence from frame $x + 1$. We can therefore say that

$$\eta_{GBN} = \frac{\overline{x}}{\overline{x} + N}\frac{k}{n}$$

Now the probability of x frames being successful followed by one that fails is $p_f (1 - p_f)^x$; therefore

$$\overline{x} = \sum_{i=0}^{\infty} ip_f(1 - p_f)^i = p_f(1 - p_f)\left[1 + 2(1 - p_f) + 3(1 - p_f)^2 + \cdots\right]$$

The sum to infinity of the series in the square brackets is $1/p_f^2$, so we find that

$$\overline{x} = \frac{1 - p_f}{p_f}$$

Hence

$$\eta_{GBN} = \frac{1 - p_f}{1 - p_f + p_f N}\frac{k}{n} \tag{1.9}$$

It may appear from this that an efficient GBN scheme would have a small value of N, however the value of N depends on the round trip delays and the frame length. Small values of N will mean long frames which in turn will have a higher error rate. In fact it is the frame error rate that is the most important term in the efficiency expression, with the factor k/n also playing its part to ensure that frames cannot be made too small.

The main difficulties with ARQ are that efficiency may be very low if the frame error rate is not kept low and that the delays are variable because they depend on the number of frame errors occurring. The delay problem may rule out ARQ for real time applications, particularly interactive ones. The solution to the efficiency problem may be to create some sort of hybrid between FEC and ARQ with FEC correcting most of the errors and reducing the frame error rate and additional error detection resulting in occasional use of the ARQ protocols.

1.14.2   Error concealment

Some applications carry data for subjective appreciation where there may still be some inherent redundancy. Examples include speech, music, images and video. In this case, the loss of a part of the data may not be subjectively important, provided that the right action is taken. Designing a concealment system is a signal processing task requiring knowledge of the application, the source coding and the subjective effects of errors. Possibilities include interpolation or extrapolation from previous values. Hybrids with FEC are also possible.

Error concealment is often appropriate for exactly the applications where ARQ is difficult or impossible. One example is digital speech where the vocoders represent filters to be applied to an input signal. The filter parameters change relatively slowly with time and so may be extrapolated when a frame contains errors. Another example occurs with music on compact disc where the system is designed in a way that errors in consecutive samples are unlikely to occur. The FEC codes have a certain amount of extra error detection and samples known to contain errors are given values interpolated from the previous and the following sample.

1.14.3   Error detection and correction capability of block codes

Error detection schemes or hybrids with FEC are usually based on block codes. In general, we can use block codes either for error detection alone, for error correction or for some combination of the two. Taking into account that we cannot correct an error that cannot be detected, we reach the following formula to determine the guaranteed error detection and correction properties, given the minimum distance of the code:

$$d_{min} > s + t \qquad\qquad (1.10)$$

where s is the number of errors to be detected and t ($\leq s$) is the number of errors to be corrected. Assuming that the sum of s and t will be the maximum possible then

$$d_{min} = s + t + 1$$

Thus if $d_{min} = 5$, the possibilities are

$$s = 4 \qquad t = 0$$
$$s = 3 \qquad t = 1$$
$$s = 2 \qquad t = 2$$

If we decided, for example, to go for single-error correction with triple-error detection, then the occurrence of four errors would be detected, but the likelihood is that the decoder would assume it was the result of a single error on a different codeword from the one transmitted.

If the code is to be used for correction of the maximum amount of errors, and if the value of minimum distance is odd, then setting $t = s$ gives

$$d_{min} = 2t + 1 \tag{1.11}$$

1.15   SELECTION OF CODING SCHEME

The factors which affect the choice of a coding scheme are the data, the channel and specific user constraints. That includes virtually everything. The data can have an effect through its structure, the nature of the information and the resulting error-rate requirements, the data rate and any real-time processing requirements. The channel affects the solution through its power and bandwidth constraints and the nature of the noise mechanisms. Specific user constraints often take the form of cost limitations, which may affect not only the codec cost but also the possibility of providing soft-decision demodulation.

1.15.1   General considerations

The major purpose of incorporating coding into the design of any system is to reduce the costs of the other components. Reliable communications can usually be obtained by simple, yet costly, methods such as increasing power. A well-designed coding scheme should result in a lower overall system cost for an equivalent or better performance. If this objective is to be met, however, the designer needs to make a careful choice and be aware of the whole range of available techniques.

Convolutional codes are highly suitable for AWGN channels, where soft decisions are relatively straightforward. The coding gains approach the asymptotic value at relatively high bit error rates, so that at bit error rates of 10^{-5} to 10^{-7} in Gaussian conditions, convolutional codes are often the best choice. Many types of conditions, however, can give rise to non-Gaussian characteristics where the soft-decision thresholds may need to adapt to the channel conditions and where the channel coherence may mean that Viterbi decoding is no longer the maximum likelihood solution. The complexity of the decoder also increases as the code rate increases above $1/2$, so that high code rates are the exception. Even at rate $1/2$, the channel speed which can be accommodated is lower than for Reed Solomon codes, although it is still possible to work at over 100 Mbits/second, which is more than enough for many applications!

Reed Solomon codes have almost exactly complementary characteristics. They do not generally use soft decisions, but their performance is best in those conditions where soft decisions are difficult, i.e. non-Gaussian conditions. In Gaussian conditions the performance curves exhibit something of a 'brick wall' characteristic, with the codes working poorly at high bit error rates but showing a sudden transition to extremely effective operation as the bit error rate reduces. Thus they may show very high asymptotic coding gains but need low bit error rates to achieve such gains. Consequently they are often advantageous when bit error rates below 10^{-10} are required. Error rates as low as this are often desirable for machine-oriented data, especially if there is no possibility of calling for a retransmission of corrupted data. The decoding complexity reduces as code rate increases, and in many cases decoding can be achieved at higher transmitted data rates. They can also, of course, be combined with other codes (including convolutional codes or other RS codes) for concatenated coding.

For the future, the so-called turbo codes are going to be of increasing importance. These are tree codes of infinite constraint length, used in combination and decoded by an iterative method. Usually two codes are used with one operating on an interleaved data set. The decoding algorithms not only use soft decisions, they also provide soft decisions on the outputs, and the output of each decoder is fed to the input of the other so that successive iterations converge on a solution. The performance is extremely good, giving acceptable error rates at values of E_b/N_0 little above the Shannon levels. There are, however, several problems to be resolved including the existence of an error floor making it difficult to achieve output BERs below 10^{-5} or 10^{-6}.

The above considerations certainly do not mean that other types of codes have no place in error control. Many considerations will lead to the adoption of other solutions, as will be seen from the discussions below. Nevertheless, mainstream interests in future systems are likely to concentrate on Viterbi-decoded convolutional codes, Reed Solomon codes and turbo codes, and the designer wishing to adopt a standard, 'off-the-shelf' solution is most likely to concentrate on these alternatives.

1.15.2   Data structure

If information is segmented into blocks, then it will fit naturally with a block coding scheme. If it can be regarded as a continuous flow, then convolutional codes will be most appropriate. For example, protecting the contents of computer memories is usually done by block coding because the system needs to be able to access limited sections of data and decode them independently of other sections. The concept of data ordering applies only over a limited span in such applications. On the other hand, a channel carrying digitized speech or television pictures might choose a convolutional scheme. The information here is considered to be a continuous stream with a definite time order. The effects of errors will be localized, but not in a way which is easy to define.

It is important to separate the structure of the data from the characteristics of the channel. The fact that a channel carries continuous data does not necessarily mean that the data is not segmented into block form. Less obvious, but equally important, a segmented transmission does not necessarily imply segmented data. A TDMA channel, for example, may concentrate several continuous streams of information into short bursts of time, but a convolutional code may still be most appropriate. With adequate buffering, the convolutional code on any stream may be continued across the time-slots imposed by the TDMA transmission.

1.15.3   Information type

It is conventional to assess the performance of coding schemes in terms that involve bit error rates. This is not really appropriate for many types of information, and the most appropriate measure will often affect the choice of a coding scheme. Indeed it is difficult to think of any application in which the bit error rate is directly important. If discrete messages are being sent, with every bit combination representing a totally different message, then the message error rate is of crucial importance; the number of bit errors in each wrong message is not important at all. Even with information that is subjected to some kind of sensory evaluation (i.e. it is intended for humans, not for machines), not all bits are equal. In most cases there are more and less significant bits or some bits whose subjective importance is different from that of others. Digitized speech without any data compression carries a number of samples, each of which has a most and a least significant bit. Only if bit errors in all positions have equal effect will bit error rate provide a measure of subjective quality. If the speech is at all compressed, the bits will represent different types of information, such as filter poles or excitation signals, and the subjective effects will vary. Data intended for subjective evaluation may be suitable for error concealment techniques.

Errors on a coded channel can be placed into four categories. There are those which are corrected by the code and allow the information to be passed on to the destination as if those errors had never occurred. There are errors which are detected but not corrected. There are also errors which are not detected at all and errors which are detected but the attempted correction gives the wrong result. Errors are passed on to the destination in the last two cases. For many applications it is important to minimize the probability of unsuspected errors in the decoder output. This will bias the user towards block codes, which often detect errors beyond the planned decoding weight, and away from forward error correction which accepts that undetected decoding errors will occur. The strength of the bias depends on the consequence of errors. If an error could start the next world war, it is obviously of more importance than one that causes a momentary crackle on a telephone line.

Acceptable error rates will depend not only on the type of data but also on whether it will be processed on- or off-line. If data is to be processed immediately, it may be possible to detect errors and invoke some other strategy such as calling for retransmission. Off-line processing means that errors cannot be detected until it is too late to do anything about it. As a result the error rate specification will commonly be lower.

Note that there must always be some level of errors which is considered to be acceptable. It is easy to set out with a goal of eliminating all errors. Achieving this goal would require infinite time and an infinite budget.

1.15.4   Data rate

It is difficult to put figures on the data rates achievable using different codes. This is partly because any figures given can quickly become out of date as technology advances and partly because greater speeds can usually be achieved by adopting a more complex, and therefore more expensive, solution. Nevertheless, for a fixed complexity, there are some codes which can be processed more rapidly than others.

The codes which can be processed at the highest data rates are essentially simple, not very powerful, codes. Examples are codes used purely for error detection. Concatenated codes using short block inner codes are not far behind because the computations on the Reed Solomon codes are done at symbol rate, not bit rate, and the block codes used are extremely simple. It follows that Reed Solomon codes alone are in the highest data rate category. Viterbi-decoded convolutional codes are fast provided the input constraint length is not too long, say no more than 9. BCH codes can also be used at similar rates provided hard-decision decoding only is required. Soft-decision decoding of block codes and the more complex concatenated schemes, e.g. turbo codes, are capable of only moderate data rates.

Of course, the required data rate affects the choice of technology too; the more that can be done in hardware the faster the decoding. Parallelism can increase decoding speeds, but with higher hardware complexity and therefore cost. A data rate of a few thousand bits per second could allow a general-purpose microprocessor to be used for a wide range of codecs, but obviously that would be uneconomic for volume production. Many of the influences of data rate on system design will be closely bound up with economics.

1.15.5   Real time data processing

If real time data processing is required, the decoder must be able to cope with the link data rates. This may be achieved at the expense of delays by, for example, decoding one sequence while the next is being buffered. The decoding delay may in some cases become significant, especially if it is variable.

Forward error correction requires a decoding delay that, in most cases, depends on the exact errors which occur. Nevertheless, there is usually a certain maximum delay that will not be exceeded. Buffering the decoded information until the maximum delay has expired can therefore produce a smooth flow of information to the destination. Two major factors determining the delay will be the data rate and the length of the code. Information theory tells us that long codes are desirable, but for many applications long delays are not. Thus the maximum acceptable delay may limit the length of the codes that can be used.

If no maximum decoding delay can be determined, then the decoded information will come through with variable delays, which can cause havoc with real time information. The main error control strategy that exhibits variable delays is ARQ because one cannot guarantee that any retransmission will be successful. These problems may be minimized by the use of a suitable ARQ/FEC hybrid.

1.15.6   Power and bandwidth constraints

These constraints drive the solution in opposite directions. In the absence of bandwidth constraints one would use a low rate concatenated code to achieve high coding gains or very low error rates. Very tight bandwidth constraints, making binary modulation incompatible with the required data rate and error rates, require the use of specially designed codes in conjunction with multilevel modulations. Traditionally these have been convolutional codes, but block codes may be possible and turbo-coded solutions are being developed.

Assuming that the major aim of coding is to reduce the power requirement for a given error rate, high coding gains would appear to be desirable. There can be no doubt that the highest gains are achievable using turbo codes or concatenated codes. If the gain requirement is less stringent, convolutional codes with hard-decision sequential decoding or soft-decision Viterbi decoding (to be described in Chapter 2) provide the highest gains on a Gaussian channel.

1.15.7   Channel error mechanisms

Ideally one would design a coding scheme for the precise conditions encountered on the channel. In practice, the channel may not be well characterized and the coding scheme may have to show flexibility to cope with the range of possible conditions. For slowly varying channel conditions which exhibit approximate Gaussian conditions over appreciable periods of time, adaptive coding schemes are a natural choice. These often use variable rate convolutional codes, or they may be based around ARQ/FEC hybrids. For channels which may fluctuate rapidly between states, producing mixtures of bursty and random errors, a wide variety of diffuse-error correcting schemes, including interleaving, are available. Reed Solomon codes may also be considered to fall into this category; although optimized neither for random errors or general bursts, their low redundancy overhead makes them a good choice for a variety of channel conditions.

1.15.8   Cost

Any error control scheme is merely a part of a larger system and its costs must be in proportion to its importance within the system. Bearing in mind that error rates may be reduced by the use of higher transmitted power, the aim of coding is to be more cost-effective than other solutions. That, however, is often not the main way in which cost constraints are experienced in a coding system; the major part of the costs of error control are incurred at the decoder, placing the burden of the economics onto the receiving equipment. Since the owners of transmitting and receiving equipment may be different, the economic considerations may not be combined to optimize overall system costs. Decoder costs must be assessed in terms of what the receiver will be prepared to pay.

A number of fairly straightforward rules may be stated. Firstly as previously indicated, the decoder costs dominate in a forward error correction scheme. Error detection is therefore much cheaper than error correction. High data rates will cost more than low data rates. Complex codes with multiple-error correction will cost more than simpler codes. For many applications, however, the main factor affecting cost will be whether there is a codec available commercially or whether it will need to be developed specially. Development costs must be spread across the number of receivers, and if the market is small or particularly cost-sensitive it may be impossible to develop a special codec for the particular needs of the application. In that case, the choice will be severely limited.

Any very specific advice about commercially available codecs would ensure that this book would quickly be out of date. As with all modern technologies, the product range is expanding and getting cheaper. Rate 1/2 Viterbi decoders are available and popular, and may incorporate puncturing for higher rates or for adaptable coding schemes (although the latter involve many other system complexities and costs). Certain Reed Solomon codes are being adopted as standard and codecs are becoming available. Often this will be a spin-off from a particular mass market, such as compact disc players.

Although it seems a shame to sound a negative note, I believe that many interesting ideas in error control will never be implemented simply because their potential market will not make the development costs worthwhile. Similarly many engineers working on error control techniques will never be allowed to design the best system technically; they will be forced to choose the best of what is available. Those who wish to have a relatively free hand should work on applications where the potential market is large or not very cost-sensitive. The same constraints apply, of course, in many other areas. Some would say that is what engineering is all about and error control is, after all, an engineering topic rather than a mathematical one. The mathematics is the servant of the engineer, and the engineering is the really difficult part.

1.16   CONCLUSION

In this chapter we have been concerned mainly with general concepts and background. There are several good modern books on digital communications that include a treatment of error control codes [2–4]. These can be consulted for more information about digital modulations implementation issues and applications. They can also be used to provide an alternative view of error control coding issues that will be treated later in this book. Another large subject given very brief treatment here is the general topic of information theory, and other sources [5, 6] are recommended for further reading.

The next chapter of this book deals with convolutional codes, which are the most commonly adopted codes in digital communications. Chapter 3 will cover linear block codes and a subset of these codes, cyclic codes, will be treated in Chapter 4. The construction of cyclic codes and the decoding methods for multiple-error correction require a knowledge of finite field arithmetic, and this is covered in Chapter 5. Chapter 6 then deals with BCH codes, a large family of binary and nonbinary codes, but concentrating on the binary examples. The most important nonbinary BCH codes are Reed Solomon codes and these are treated in Chapter 7. Chapter 8 then deals with performance issues relevant to all block codes. Multistage coding is introduced in chapter 9. Codes using soft-in-soft-out algorithms for iterative decoding are covered in Chapter 10.

It should not be assumed that the length of the treatment of different codes indicates their relative importance. Block codes have a very strong theoretical basis, but for many applications the chapters on convolutional codes and on iterative decoding will be the most relevant. Iterative decoding can, however, be applied to block codes and familiarity with Chapters 4 and 9 will certainly help in obtaining the most from Chapter 10.

1.17 EXERCISES

1   A transmitter sends one of four possible messages. The most likely occurs with a probability of 0.5 and is represented by the value 0. The second most likely occurs with a probability of 0.25 and is given the representation 10. The other two messages each occur with a probability of 0.125 and are given the representations 110 and 111. Find the mean information content of the messages and the mean number of bits transmitted. Compare with the case where a fixed length representation is used.

2   Find the value of E_r/N_0 needed by BPSK or QPSK modulation to achieve a bit error rate of 10^{-3} over an AWGN channel.

3   Find approximate optimum uniform spacing for 16-level quantization from a BPSK or QPSK receiver, assuming an AWGN channel with $E_r/N_0 = 2\,\text{dB}$.

4   Use Table 1.1 to carry out minimum distance decoding of hard-decision sequences 01011, 01110, 10100, 11000. Use the method of Section 1.8.3 to check the results.

5   Prove that, for a linear code, the distance structure is the same viewed from any codeword.

6   An uncoded channel needs E_b/N_0 of 2 dB to achieve a BER of 10^{-2} and 10 dB to achieve a BER of 10^{-5}. A rate 1/2 code subject to random bit errors with probability of 0.01 produces an output BER of 10^{-5}. What is the coding gain at 10^{-5} BER? If the coded channel operates at $E_b/N_0 = 5\,\text{dB}$, what is the uncoded BER?

7   Find the maximum coding gain that could be achieved for BER $= 10^{-6}$ using QPSK modulation over an AWGN channel and a rate 1/3 code.

8   A block code has $d_{min} = 8$. Find the maximum guaranteed error detection if maximum error correction is to be carried out. How would this change if only single-error patterns were to be corrected? Find the amount of error correction achievable if an *extra* three bits in error should be detectable.

9   A source transmits a sequence of numbered frames, inserting repeats of previous frames as required for ARQ. The second, sixth and ninth frames transmitted are corrupted on reception. Show the flow of frames between the transmitter and receiver to support a GB3 protocol.

10   Assuming the same propagation delays and the same corrupted frames as question 9, find how many frame intervals would be needed to transfer 7 frames if SR-ARQ were used, assuming

(a)   an infinite receiver window
(b)   a minimum receiver window to operate SR-ARQ.

Quantify the receiver window size needed for part (b).

1.18   REFERENCES

1   J. Massey, *Coding and Modulation in Digital Communications*, Proceedings of the International Zurich Seminar on Digital Communications, 1984.
2   S. Haykin, *Communication Systems*, 4th edition, John Wiley & Sons, 2001.
3   B. Sklar, *Digital Communications: fundamentals and applications*, 2nd edition, Prentice Hall, 2001.
4   J.G. Proakis, *Digital Communications*, 4th edition, McGraw Hill, 2001.
5   S.W. Golomb, R.E. Peile and R.A. Scholtz, *Basic Concepts in Information Theory and Coding*, Plenum Press, 1994.
6   R.E. Blahut, *Principles and Practice of Information Theory*, Addison Wesley, 1987.

2

Convolutional codes

2.1 INTRODUCTION

In Chapter 1 it was explained that codes for error control generally fell into two categories, namely block codes and convolutional codes. Many telecommunications applications have used convolutional codes because of their ability to deliver good coding gains on the AWGN channel for target bit error rates around 10^{-5}.

As background to this chapter most of the first chapter is relevant, even though the examples in Chapter 1 were based on a block code. The main decoding method, Viterbi decoding, is indeed an implementation of the maximum likelihood decoding that was explained there. This means that the discussion on soft-decision decoding will be relevant as it will almost always be implemented in conjunction with Viterbi decoding. Convolutional codes are also linear, as previously defined.

2.2 GENERAL PROPERTIES OF CONVOLUTIONAL CODES

An example of the schematic diagram of a convolutional encoder is shown in Figure 2.1. Note that the information bits do not flow through directly into the code-stream, i.e. the code is not systematic. Nonsystematic codes give better performance than systematic codes when Viterbi decoding is used.

The way in which the encoder works is that the input bit is modulo-2 added to stored values of previous input bits, as shown in the diagram, to form the outputs

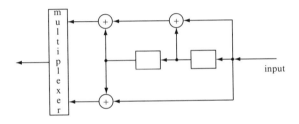

Figure 2.1 Convolutional encoder

which are buffered ready for transmission. The input bit is then moved into the shift registers and all the other bits shift to the left (the leftmost, i.e. oldest, stored bit being lost).

We can construct a truth table for the encoder as follows. If the state of the encoder is taken as the content of the registers, with the oldest bit to the left and the most recent to the right, then the possible cases are shown in Table 2.1.

The output bits are listed with the output from the upper encoder branch to the left of the bit from the lower branch. Note that the value of the rightmost bit of the end state is the same as the input bit because the input bit will be stored in the rightmost register.

Consider as an example the encoding of the sequence 1 0 1 1 0 1 0. The encoder will start with the registers clear, i.e. from state 00 and the encoding operation will proceed as shown in Table 2.2.

The start state for any operation can easily be found from the previous two inputs. Note that at the end of the sequence the encoder has been cleared by applying two more zeros to the input. This is because the decoding will be more robust if the encoder finishes in, as well as starts from, a known state.

Table 2.1   Truth table for convolutional encoder

Start state	Input	End state	Output
00	0	00	00
00	1	01	11
01	0	10	10
01	1	01	01
10	0	00	11
10	1	01	00
11	0	10	01
11	1	11	10

Table 2.2   Encoding example

Input	1	0	1	1	0	1	0	(0)	(0)
Output	11	10	00	01	01	00	10	11	00

2.3   GENERATOR POLYNOMIALS

The encoding operation can be described by two polynomials, one to represent the generation of each output bit from the input bit. For the above code they are

$$g^{(1)}(D) = D^2 + D + 1$$
$$g^{(0)}(D) = D^2 + 1$$

The operator D represents a single frame delay.

The interpretation of these polynomials is that the first output bit is given by the modulo-2 sum of the bit that has been remembered for two frames (the D^2 term), the bit remembered for one frame (D) and the input bit (1). The second output is the bit remembered for two frames (D^2) modulo-2 added to the input (1). This can be seen to correspond to Figure 2.1. Generator polynomials for good convolutional codes are often tabulated in the literature, represented in octal or in hexadecimal form. The first polynomial above has coefficients 1 1 1, represented as 7, and the second polynomial has coefficients 1 0 1, represented as 5.

The concept of generator polynomials can be applied also to cases where several bits are input at once. There would then be a generator to describe the way that each of the input bits and its previous values affected each of the outputs. For example a code which had two bits in and three out would need six generators designated $g_1^{(2)}(D)$, $g_1^{(1)}(D)$, $g_1^{(0)}(D)$, $g_0^{(2)}(D)$, $g_0^{(1)}(D)$ and $g_0^{(0)}(D)$.

2.4   TERMINOLOGY

The terms to be used here to describe a convolutional code are as follows:

- Input frame – the number of bits, k_0, taken into the encoder at once.

- Output frame – the number of bits, n_0, output from the encoder at once.

- Memory order – the maximum number, m, of shift register stages in the path to any output bit.

- Memory constraint length – the total number, v, of shift register stages in the encoder, excluding any buffering of input and output frames.

- Input constraint length – the total number, K, of bits involved in the encoding operation, equal to $v + k_0$.

A term which may cause particular problems in the literature is *constraint length*. Used without qualification, it most often means what I have called *input constraint length*, but it could also be the *memory constraint length*. A convolutional code may be termed a (n_0, k_0, m) code; however, the value of k_0 is almost always equal to 1 and so the codes are most commonly designated by their rate, k_0/n_0, and their constraint length (however defined).

For our example, $k_0 = 1$, $n_0 = 2$, $m = 2$, $v = 2$, and $K = 3$. The code is a (2, 1, 2) convolutional code, or a rate 1/2 code with input constraint length 3.

Convolutional codes are part of a larger family of codes called tree codes, which may be nonlinear and have infinite constraint length. If a tree code has finite constraint length then it is a trellis code. A linear trellis code is a convolutional code.

2.5   ENCODER STATE DIAGRAM

If an encoder has v shift register stages, then the contents of those shift registers can take 2^v states. The way in which the encoder transits between states will depend on the inputs presented in each frame. The number of possible input permutations to the encoder in a single frame is 2^{k_0}. Hence if $v > k_0$ not all states can be reached in the course of a single frame, with only certain states being connected by allowed transitions. This can be seen in the example case from Table 2.1.

The encoder states can be represented in diagrammatic form with arcs to show allowed transitions and the associated input and output frames, as in Figure 2.2 which shows the transitions for the encoder of Figure 2.1. As in the truth table, the states are labelled according to the contents of the encoder memory and the bits in the output frames are listed with the bit out of the upper branch shown to the left of the bit from the lower branch. Since the input frame value is represented in the end state number, it is not always necessary to include input values in the state diagram.

The memory constraint length is easily determined from the number of states. The number of branches from each state (and the labelling if inputs are shown) allows the size of the input frame to be determined. The shortest path from the zero state (memory clear) to the state where the memory is filled with ones gives the memory order.

It is also fairly easy to work in reverse order and to derive the encoder circuit or the generator polynomials from the state diagram. To find the generator polynomials describing the output contribution of a single bit of the input frame, we start with the encoder clear and apply a 1 to the input. The corresponding outputs show whether the input is connected to that output branch, i.e. they give the unity terms in the generator polynomials. Next we apply a zero to the input so that the outputs show whether the stored value from the previous frame is connected to the corresponding output branch, i.e. they give the coefficients of D in the generator polynomials. Similarly, applying further zeros to the input will give the coefficients of the higher powers of D.

For our example, the first output frame would be 11, indicating that the unity term is present in both polynomials. The second output frame of 10 indicates that the D term is present in $g^{(1)}(D)$ but not in $g^{(0)}(D)$. The third output frame of 11 indicates that both polynomials contain the D^2 term.

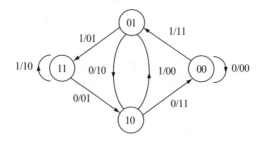

Figure 2.2   Encoder state diagram

2.6 DISTANCE STRUCTURE OF CONVOLUTIONAL CODES

As with block codes, there are concepts of distance which determine the error-correcting properties of the code. Because of linearity, one can assess the distance properties of the code relative to the all-zero sequence but, since the code sequence may be infinite in length, it is not at all clear how long the sequences to be compared should be. In particular, one might think that there will be an infinite number of differences between two different sequences of infinite length. However, this is not the case because two infinite sequences can differ over a finite length of time.

We want to compare two paths which start from the same point and then diverge but later converge. Linearity means that we can use any path as the baseline for our comparison, and as in the block code case, the all-zero sequence is convenient. We will therefore look for a code path which leaves the zero state, returning to it some time later and in the process producing a minimum number of 1s on the output. In other words, we want the lowest weight path that leaves state zero and returns to that state. Such paths are known as *free paths*. Looking at the state diagram, there are really only two sensible paths to consider. The path connecting states 00–01–10–00 results in output frames 11 10 11, Hamming weight 5. The path connecting states 00–01–11–10–00 results in output frames 11 01 01 11, Hamming weight 6. The minimum weight is therefore 5 and we term this the *free distance* (d_{free} or d_∞) of the code.

2.7 EVALUATING DISTANCE AND WEIGHT STRUCTURES

The performance of convolutional codes depends not only on the free distance of the code but also on the number of paths of different output weights and the weight of the input sequences giving rise to those paths. It is therefore important to evaluate accurately the important paths through the state diagram. The method is explained in the context of the previous example, based on the state diagram of Figure 2.2. The start point is to rearrange the state diagram so that the state 00 appears at each end of a network of paths, thus representing both the start and end points of the paths of interest. This is shown in Figure 2.3.

As the encoder moves from state to state, three things of possible interest happen. The length of the code sequence increases and the weights of both the input and output sequences either increase or remain the same. We define operators W, corresponding to an increase of 1 in the output weight, L, corresponding to an increase of one frame in the code sequence length and I, representing an increase of 1 in the input sequence weight. We can now label each arc of the modified state diagram with the appropriate operators, as has been done in Figure 2.3.

We let X_i represent the accumulated weights and lengths associated with state i and multiply the initial values by the values against the arcs to represent final state values. Of course each state can be entered from more than one start state, but if we show the sums of the contributions from each of the possible start states, all the possible paths

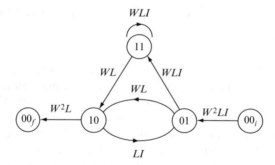

Figure 2.3   Modified encoder state diagram

through the encoder will appear as separate terms in the final expression. We can thus set up a number of simultaneous equations representing the different states as follows:

$$X_{01} = W^2 LI\, X_{00_i} + LI\, X_{10} \tag{2.1}$$

$$X_{10} = WL\, X_{01} + WL\, X_{11} \tag{2.2}$$

$$X_{11} = WLI\, X_{01} + WLI\, X_{11} \tag{2.3}$$

$$X_{00_f} = W^2 L\, X_{10} \tag{2.4}$$

We want to know what happens when moving from state 00_i to 00_f, so we divide X_{00_f} by X_{00_i} to yield the input and output weights and the lengths of all the possible paths. This is achieved as follows. From (2.3) we obtain

$$X_{11} = X_{01}\frac{WLI}{1 - WLI}$$

Substituting into (2.2) gives

$$X_{10} = X_{01} WL\left(1 + \frac{WLI}{1 - WLI}\right) X_{01}\frac{WL}{1 - WLI}$$

Now eliminate X_{01} from (2.1)

$$X_{10}\left(\frac{1 - WLI}{WL} - LI\right) = X_{00_i} W^2 LI$$

$$X_{10}(1 - WLI - WL^2 I) = X_{00_i} W^3 L^2 I$$

$$X_{10} = X_{00_i}\frac{W^3 L^2 I}{[1 - WLI(1 + L)]}$$

Finally, substituting into (2.4) gives

$$\frac{X_{00_f}}{X_{00_i}} = \frac{W^5 L^3 I}{1 - WLI(1 + L)}$$

A binomial expansion on this expression gives

$$\frac{X_{00_f}}{X_{00_i}} = W^5 L^3 I \left[1 + WLI(1 + L) + W^2 L^2 N^2 (1 + L)^2 + \cdots \right]$$

$$\frac{X_{00_f}}{X_{00_i}} = W^5 L^3 I + W^6 L^4 I^2 + W^6 L^5 I^2 + W^7 L^5 I^3 + 2 W^7 L^6 I^3 + W^7 L^7 I^3 \cdots$$

This tells us that between states 00 and 00 there is one path of length 3, output weight 5 and input weight 1; a path of length 4, output weight 6 and input weight 2; a path of length 5, output weight 6 and input weight 2; a path of length 5, output weight 7 and input weight 3; two paths of length 6, output weight 7 and input weight 3; a path of length 7, output weight 7 and input weight 3, etc.

The expression for X_{00_f}/X_{00_i} is called the *generating function* or the *transfer function* of the encoder. It will be used to find the performance of convolutional codes with maximum likelihood decoding in Section 2.12.

2.8   MAXIMUM LIKELIHOOD DECODING

In principle the best way of decoding against random errors is to compare the received sequence with every possible code sequence. This process is best envisaged using a code trellis which contains the information of the state diagram, but also uses time as a horizontal axis to show the possible paths through the states. Code trellises get very complex for large constraint lengths and so we shall take just one example, shown in Figure 2.4, for the encoder of Figure 2.1. The encoder states are shown on the left and the lines, moving right to left, show the allowed state transitions. The labels against each transition are the encoder outputs associated with each transition. As for the state diagram, the inputs to the encoder have not been shown as they can be deduced from the end state.

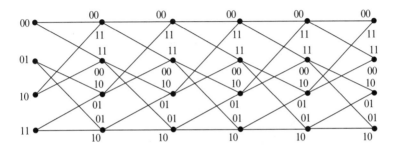

Figure 2.4   Code trellis

The apparent problem with maximum likelihood decoding is that over L code frames there are 2^{Lk_0} paths through the trellis, and comparing what is received with every possible path seems unmanageable. Fortunately, Viterbi realized that not all of these paths need be considered, and that at any stage only 2^v paths need to be retained, provided the errors show no correlation between frames (memoryless channel). He devised a technique which simplifies the problem of decoding without sacrificing any of the code's properties.

2.9   VITERBI ALGORITHM

2.9.1   General principles

If we look at all the paths going through a single node in the trellis and consider only the part from the start of transmission up to the selected node, we can compute the distance between the received sequence and each of these trellis paths. When we consider these distance metrics we will probably find that one of the paths is better than all the others. Viterbi realized that if the channel errors are random then the paths which are nonoptimal at this stage can never become optimal in the future. In other words, we need keep only one of the paths reaching each node. The Viterbi method therefore keeps only 2^v paths through the trellis and at each frame it decides which paths to keep and which to discard. The procedure at each received frame is:

(a)   For each of the 2^v stored paths at the start of the frame, compute distance between the received frame and the 2^{k_0} branches extending that path. This is called a *branch metric*.

(b)   For each of the 2^v nodes which represent the end states of the frame, construct the 2^{k_0} paths which terminate at that node. For each of those paths, evaluate the sum of branch metrics from the start of the sequence to reach an overall path metric. Note that the path metric can be computed as the sum of the branch metric and the previous path metric. Select and store the best path.

2.9.2   Example of Viterbi decoding

Consider an example based on Figure 2.4. Let us assume that we receive the sequence 11 10 10 01 11. In the first frame, the computation of branch and path metrics is as shown in Figure 2.5. In this frame, only the paths originating from state 00 have been considered. The received sequence has a Hamming distance of 2 to the sequence (00) that would be transmitted if the encoder stays in state 00 and a Hamming distance of 0 to the sequence (11) that would be transmitted if the encoder transits from state 00 to state 01. As this is the start of the example, the branch metrics, shown in parentheses, and the path metrics, shown against each possible end state, are the same.

In the second frame, the Hamming distance from the received sequence 10 is computed for each of the four possible branches as shown in Figure 2.6. The path metric to each end state is computed as the branch metric plus the previous path metric.

The decoding of the third frame, for which the sequence 10 is received, is shown in Figure 2.7. Firstly the eight possible branch metrics have been computed and are shown in parentheses. Next the path metrics are computed as the sum of the branch metric plus the previous path metric. From this point, however, it is seen that there are two possible ways to reach each end node and so both path metrics are computed, with the lower value and the corresponding path being retained. The retained paths have been shown in bold in the figure.

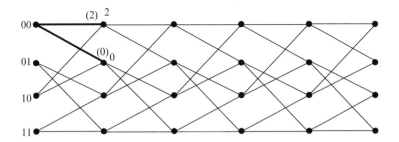

Figure 2.5   Metrics for received 11 on the first frame

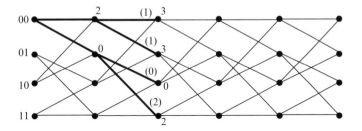

Figure 2.6   Metrics for received 10 on the second frame

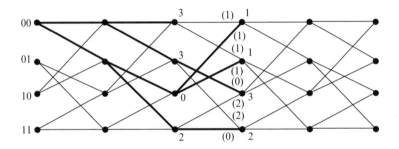

Figure 2.7   Metrics for received 10 on the third frame

In the fourth frame, the received sequence is 01. The branch metrics are different from those of the previous frame, but otherwise the processing of this frame is similar to that of the previous frame. This is shown in Figure 2.8.

In the final frame of our example, the received sequence is 11 and the result of the branch and path metric computation is shown in Figure 2.9.

Our example is now completed, but it is by no means clear what the solution is as there are four retained paths. Remember, however, that it was stated earlier that we would clear the encoder at the end of the data. Assuming that has happened, the encoder must end in state 00. Therefore, provided we know that the end of the transmitted sequence has been reached, we choose the path that finishes in state 00. This is true, regardless of the path metrics at the end.

The path ending in state 00 is seen to have gone through the state transitions 00–01–11–11–10–00. The first transition (00–01) must have been caused by a 1 on the encoder input since the final state ends in 1. Similarly the second (01–11) and third (11–11) transitions must also have resulted from a 1 at the encoder input. The fourth (11–10) and fifth (10–00) transitions result from a 0 at the encoder input. The complete decoded sequence is therefore 1 1 1 0 0. The two final zeros, however, will have been the flushing or clearing bits, not real data. The decoded data is therefore 1 1 1.

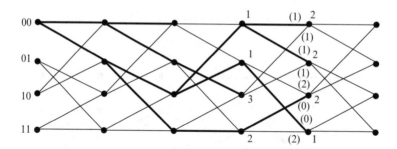

Figure 2.8 Metrics for received 01 on the fourth frame

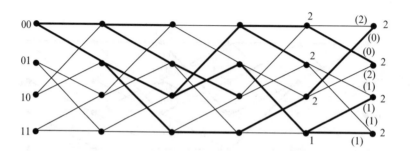

Figure 2.9 Metrics for received 11 on the fifth frame

2.9.3   Issues arising

There are some issues that will arise in practical implementations of Viterbi decoding that were not addressed in the above example, but can be dealt with quickly.

One obvious problem is that two paths might merge with the same metric. In that case, it will be necessary to make an arbitrary choice between them. Nothing in the future will make that choice easier, so there is no point in deferring the decision. Looking at Figure 2.9, it can be seen that many of the paths through the trellis die out eventually, so there is at least a chance that the arbitrary choice may not figure in the final solution.

We have already stated that, at the end of the example, we choose the path ending in state 00, because we know that the encoder will be returned to that state at the end of the sequence. In practice, however, the transmitted sequence may be very long and waiting until the end may not be acceptable. Looking again at Figure 2.9, we can see that all of the paths following the transition 00–00 in the first frame have died out by the end of the fourth frame. The path 00–00–00 is not retained at the end of the third frame. The path 00–00–01 is extended further to state 10 in the third frame, but neither of its extensions survives the fourth frame. Therefore all the paths being retained from frame 4 onwards start with the transition 00–01. As a result, we know after the fourth frame that the first data bit will be decoded as 1. There is therefore a possibility that we can commence decoding before the end of the sequence when all retained paths agree in the early frames. This point will be further examined in the next section, along with matters such as soft-decision decoding and organization of the path and metric storage.

2.10   PRACTICAL IMPLEMENTATION OF VITERBI DECODING

In this section we shall see an example that will extend on the previous one to incorporate several other practical features. Firstly, we shall organize the path storage in a reasonably efficient way. Secondly, soft-decision decoding will be implemented. In addition, certain other practical features will also be discussed.

We need to consider the organization of the path and metric storage. We would like to be able to identify the path finishing at each end state without the need to inspect the stored values; in other words, we would like a particular storage location to be associated with a particular end state. However, when we extend a path it may then finish in a different state, but we do not want to have to swap paths between locations. Obviously an extended path could overwrite the path in the appropriate location, but not until the overwritten path itself has been extended. We also want to minimize the amount of storage needed.

One possible solution is to record at each node the identity of the previous node in the survivor path, as shown in the above example. There will need to be a separate circuit or routine to trace back the desired path to complete the decoding and, if this is to be done concurrently with processing of received frames, the stored paths will continue to grow during the trace back operations. However there will be no need to manipulate the contents of registers.

Another possible solution is to have two sets of registers, a set A associated with the start states at the beginning of the frame and a set B associated with the end states after the path extensions. For each location in the B registers, we create the two paths that merge at the corresponding location and store the one with the better path metric. When all path extensions are finished, we swap the pointers associated with the set A and the set B so that the end states for one frame become the start states for the next and the start states for the previous frame become irrelevant and can be overwritten. For this approach there is no need to trace back any path because the entire path is held in a single register. There is a need, however, to copy the contents of several registers with each new frame.

The solution to be adopted here will be similar to this second approach. There will again be two registers associated with each end state, but these will be used to hold all the path extensions. The values will then be compared in pairs and the good path will overwrite the less good one. For our example, denote the registers A00, A01, A10, A11, B00, B01, B10, B11. Only the A registers are used at the start of the frame. The extensions of the path in register A00 will be written in B00 and B01. The extensions from register A01 will be stored in registers B10 and B11. The path in register A10 will extend to locations A00 and A01. The path in A11 will extend to A10 and A11. Finally both the A and B registers for a given end state will be compared, the better metric chosen and the good path written into the A register for subsequent extension.

As stated in Chapter 1, received bits could possibly be given soft-decision values from 0 to 7 and distance computed on the basis that transmitted bits have a numerical value of 0 or 7. For this example, however, we shall use a correlation metric which effectively measures similarity between the received sequence and the code sequence and can take positive or negative values. This is the metric that was adopted for the block code soft-decision decoding example in Table 1.16.

Maintaining the 8-level or 3-bit quantization of the received bits, the levels will be interpreted as having values 3.5, 2.5, 1.5, 0.5, −0.5, −1.5, −2.5, −3.5. The path metric associated with a received level r_i will be $+r_i$ for a transmitted 1 and $-r_i$ for a transmitted 0 and therefore for a received frame the branch metrics will be in the range −7 to +7 in steps of 1. As an example, we shall consider the received sequence +3.5 +1.5, +2.5 − 3.5, +2.5 − 3.5, −3.5 +3.5, +2.5 + 3.5.

As we extend the paths, the metrics will grow. However the difference between the maximum and minimum metric is bounded and we can always obtain a unique value for any metric difference provided the worst case range of the metrics is less than one half the range allowed by the storage registers. In this example, an 8-bit storage register is assumed, representing values from +127 to −128 in steps of 1. Also, to keep the operations the same for every frame, we load the path metrics for all nonzero states with a large negative value (taken as −64) at the start of the decoding to ensure that only the path starting from state zero survives the first few frames.

At the start of the example, the register contents are as shown in Table 2.3. The B registers are not used as the start point for path extensions, so their content is not relevant.

In the first frame, the received sequence is +3.5 +1.5. The path metrics are now as shown in Table 2.4, with the metrics in locations B00 and B01 representing the extensions from state 00, and all the others being produced as a result of the initial loading on nonzero states.

Now comparing in pairs and writing the higher metric path into the A register give the result shown in Table 2.5. The B register content is not shown as it is not used in subsequent stages.

Table 2.3 Initialization of stored paths and metrics

Location	Path	Metric
A00		0
A01		−64
A10		−64
A11		−64
B00		
B01		
B10		
B11		

Table 2.4 Stored paths and metrics after first frame extensions

Location	Path	Metric
A00	0	−69
A01	1	−59
A10	0	−66
A11	1	−62
B00	0	−5
B01	1	+5
B10	0	−62
B11	1	−66

Table 2.5 Stored paths and metrics after first frame path merging

Location	Path	Metric
A00	0	−5
A01	1	+5
A10	0	−62
A11	1	−62
B00		
B01		
B10		
B11		

In the second frame, the received sequence is $+2.5, -3.5$, producing path metrics shown in Table 2.6.

Writing the good path of each pair into the A register gives the values shown in Table 2.7.

In the third frame, the received sequence is $+2.5 -3.5$. The path metrics are shown in Table 2.8.

Writing the good path of each pair into the A register gives the results shown in Table 2.9.

Table 2.6 Stored paths and metrics after second frame extensions

Location	Path	Metric
A00	00	−63
A01	01	−61
A10	10	−68
A11	11	−56
B00	00	−4
B01	01	−6
B10	10	+11
B11	11	−1

Table 2.7 Stored paths and metrics after second frame path merging

Location	Path	Metric
A00	00	−4
A01	01	−6
A10	10	+11
A11	11	−1
B00		
B01		
B10		
B11		

Table 2.8 Stored paths and metrics after third frame extensions

Location	Path	Metric
A00	100	+10
A01	101	+12
A10	110	−7
A11	111	+5
B00	000	−3
B01	001	−5
B10	010	0
B11	011	−12

In the fourth frame, the received sequence is −3.5 +3.5. The path metrics are shown in Table 2.10.

Writing the good path of each pair into the A register gives the result shown in Table 2.11.

In the fifth frame of the example, the received sequence is +2.5 +3.5. The path metrics are shown in Table 2.12.

Table 2.9 Stored paths and metrics after third frame path merging

Location	Path	Metric
A00	100	+10
A01	101	+12
A10	010	0
A11	111	+5
B00		
B01		
B10		
B11		

Table 2.10 Stored paths and metrics after fourth frame extensions

Location	Path	Metric
A00	0100	0
A01	0101	0
A10	1110	+12
A11	1111	−2
B00	1000	+10
B01	1001	+10
B10	1010	+5
B11	1011	+19

Table 2.11 Stored paths and metrics after fourth frame path merging

Location	Path	Metric
A00	1000	+10
A01	1001	+10
A10	1110	+12
A11	1011	+19
B00		
B01		
B10		
B11		

Table 2.12 Stored paths and metrics after fifth frame extensions

Location	Path	Metric
A00	11100	+18
A01	11101	+6
A10	10110	+20
A11	10111	+18
B00	10000	+4
B01	10001	+16
B10	10010	+9
B11	10011	+11

Table 2.13 Stored paths and metrics after fifth frame path merging

Location	Path	Metric
A00	11100	+18
A01	10001	+16
A10	10110	+20
A11	10111	+18
B00		
B01		
B10		
B11		

Writing the good path of each pair into the A register gives the result shown in Table 2.13.

The example concludes by selecting the path that finishes in state 00, i.e. the path in register A00, even though it is not the path with the best metric. As in the hard-decision example, the decoded sequence is 11100, of which the last two zeros are the clearing bits.

Note that from the end of frame four all the registers contained paths starting with the data bit value 1. This bit could have been decoded at that point and there would be no possibility of subsequent change. In practice deciding by inspection when to decode a particular input frame would be time-consuming, so either decoding is carried out at the end of the sequence or else a fixed delay is imposed, usually of the order of four or five memory orders. This is the typical upper limit to the length of a decoding error event, after which two trellis paths that have diverged will have merged together again. By maintaining a path length of this order, there is a high likelihood that the stored paths will all agree at the beginning so that the probability of error resulting from a lack of path agreement is small compared with the usual values of probability of decoding error.

For this example code, it might be decided to allocate one byte (eight bits) to each path storage location so that after eight frames the locations are full. The first bit is taken from one of the registers as the data bit for the first frame and all the registers are then shifted left to leave a free location for the next frame. Decoding for each subsequent frame proceeds in the same way until the end of the sequence is reached, when the entire decoding can be completed from register A00. Before the end, it does not matter which register is used for each decoded data bit since all registers are assumed to agree at the beginning, but in practice it is easiest to stick with register A00. Alternatively, it might be decided to use the start of the path with the best metric for the decoded data.

The decoding delay to be implemented can have a measureable effect on decoded error rates. For any finite register size, there is always a finite possibility that the stored paths will not agree when the registers are filled. As register size is increased, the measured error rates for a given channel condition will reduce, but a law of diminishing returns will apply. The point where this happens depends on the inherent error rates of the channel–code combination, but convolutional coding schemes are often designed to deliver bit error rates of 10^{-5} or perhaps 10^{-6}. At these levels, a register length of four to five memory orders is found to cause negligible degradation to the performance, compared with decoding at the end of the sequence.

2.11   PERFORMANCE OF CONVOLUTIONAL CODES

The calculation of performance of a convolutional code will draw on the transfer function of the code as described in Section 2.7. It will use a technique known as a *Union Bound* calculation. The terminology derives from the fact that the size of the union of two sets is at most the sum of the sizes of the individual sets. The principle can be understood as follows.

Suppose we roll an unbiased dice and are interested in the possibility of obtaining a 6. The probability of this is just $1/6$. Now suppose we are allowed two throws to obtain a 6; the probability of success is now $2/6$, right? Wrong! By that logic, with six throws there would be a probability 1 of success, but it is obvious that is not the case. We are allowed to add probabilities only when the events described are mutually exclusive, i.e. there is no chance that both will occur. In our example, we could get a 6 both times and the probability of that happening is exactly the overestimate of probability of success.

More generally, for two independent events 1 and 2 which occur with probabilities p_1 and p_2, respectively, the probability that either event will occur can be calculated as 1 minus the probability that neither event will occur:

$$p(1 \text{ and } 2) = 1 - (1 - p_1) \cdot (1 - p_2) = p_1 + p_2 - p_1 \cdot p_2$$

We could also phrase this as

$$p(1 \text{ and } 2) \leq p_1 + p_2$$

where the equality would apply if the events are mutually exclusive. This is exactly the form of the Union Bound computations in which we shall express the probability of at least one of a number of events occurring as being upper bounded by the sum of the individual probabilities.

Now let us consider the issue of decoding errors. We assume that the transmitted sequence was the all-zero sequence, but that the noise on the channel causes the decoder to choose a different sequence which leaves state zero and later returns to it. This sequence will be of weight w and will correspond to a data sequence of weight i. The incorrect choice of this sequence will therefore result in i bit errors at the output of the decoder.

When considering the probability of incorrectly choosing a particular weight w path, it can be seen that it does not depend on the location of the ones in the code sequence or on how widely they are separated. In choosing between two paths, we need consider only the positions in which they differ and we could, without loss of generality, compare a path of w zeros with a path of w ones. The probability of that particular error therefore depends only on the weight. We shall denote this probability $p(w)$.

Suppose there are $T_{w,i}$ paths of weight w and input weight i in the code. The probability that a particular input frame will correspond to the start (or the end) of a decoding error is, using the Union Bound calculation,

$$p_{de} \leq \sum_w \sum_i T_{w,i} p(w) \tag{2.5}$$

Each of the events being considered will be associated with i bit errors and calculation of the output bit error rate must be normalized to k_0, the number of input bits in the frame:

$$BER \leq \frac{1}{k_0} \sum_w \sum_i i T_{w,i} p(w) \tag{2.6}$$

The procedure for calculating output bit error rates is therefore as follows:

From the transfer function, eliminate L (by setting $L = 1$) to find $T(W, I) = \sum_w \sum_i T_{w,i} W^w I^i$.

For each w, calculate $A_w = \sum_i i T_{w,i}$.

For each w, evaluate $p(w)$ (as shown below).

Find

$$BER \leq \frac{1}{k_0} \sum_w A_w p(w) \tag{2.7}$$

For our example code, we had

$$T(W, L, I) = W^5 L^3 I + W^6 L^4 I^2 + W^6 L^5 I^2 + W^7 L^5 I^3 + 2W^7 L^6 I^3 + W^7 L^7 I^3 \cdots$$

Eliminating L gives $T(W, I) = W^5 I + 2W^6 I^2 + 4W^7 I^3 + \cdots$ from which we find that $A_5 = 1$, $A_6 = 4$ and $A_7 = 12$. Although these coefficients are growing with higher values of w, the value of $p(w)$ usually declines rapidly with increasing w so that only the first few terms are generally needed.

To find $p(w)$ for unquantized soft-decision demodulation of binary PSK (or QPSK) on the AWGN channel, recall that the uncoded BER is $p = 1/2\, erfc\sqrt{E_r/N_0}$, where E_r is the energy per received bit and N_0 is the single-sided noise power spectral density. When using a code of rate R, the value of E_r is equal to RE_b, where E_b is the energy per bit of information. However, the decoder will be able to consider w bits in discriminating between sequences, so that the signal energy involved in the comparisons will effectively be increased by a factor of w. As a result, we obtain

$$p(w) = \frac{1}{2} erfc\sqrt{wR\frac{E_b}{N_0}} \qquad (2.8)$$

This is substituted in Equation (2.7).

Using 8-level (3-bit) quantization of the soft decisions degrades the performance by around 0.25 dB.

For hard-decision demodulation, the value of $p(w)$ is found to be

$$p(w) = 2\sqrt{p(1-p)} \qquad (2.9)$$

where p is the bit error rate from the demodulator.

2.12   GOOD CONVOLUTIONAL CODES

Assuming Viterbi decoding, the memory order of convolutional codes is likely to be in single figures. At every frame received, the decoder has to update 2^v states and for each of these states there are 2^{k_0} paths to be evaluated. Thus the amount of computation in the decoder is roughly proportional to $2^{v+k_0} = 2^K$. This sets an upper limit to constraint lengths of the codes which can be decoded in this way. The limit depends on technology and required bit rate, but figures of $K = 7$ to 9 are commonly quoted as typical present day maxima. Larger constraint lengths, which mean more powerful codes, can only be decoded at reasonable rates by other techniques such as sequential decoding to be described in Section 2.16.

The generators, in octal form, of some known good rate 1/2 convolutional codes are shown in Table 2.14.

Table 2.14   Rate 1/2 convolutional codes

m	$g^{(1)}, g^{(0)}$	d_∞
2	7, 5	5
3	17, 15	6
4	35, 23	7
5	75, 53	8
6	171, 133	10
7	371, 247	10
8	753, 561	12

The most commonly encountered convolutional code is rate $1/2$ and has input constraint length $K = 7$ and generators

$$g^{(1)}(D) = D^6 + D^5 + D^4 + D^3 + 1$$

$$g^{(0)}(D) = D^6 + D^4 + D^3 + D + 1$$

The coefficients used to calculate output bit error rate for this code are $A_{10} = 36$, $A_{12} = 211$, $A_{14} = 1404$, $A_{160} = 11\,633$.

Finding the above good convolutional codes was done by computer search based on methods that we have now encountered. For a particular rate and constraint length, all possible generator polynomials can be constructed and the transfer function of the resulting code can be generated by computerized methods. Then the Union Bound performance can be plotted. Obviously one does not wish to go through the entire process for codes that are definitely not optimum, and to enable some screening, the free distance will give a first indication of a code's likely performance. Nonsystematic convolutional codes allow greater values of d_{free} to be obtained than for systematic codes. Reversing the order of a set of connections from the storage registers will not affect the code's properties, so certain possibilities can be grouped together. In addition, some codes will be eliminated on the basis that they exhibit *catastrophic error propagation*, as explained below.

Considering the state diagram of a code, suppose that another state, apart from state zero, had a zero-weight loop returning to it. It would now be possible to devise a code sequence which starts from the zero state and ends with a sequence of zeros, but in which the encoder has not returned to the zero state. Moreover, since the encoder state is the result of recent inputs, it cannot be a sequence of input zeros that is maintaining the encoder in that nonzero state. Thus comparing this sequence with the all-zero sequence, we could have two information sequences which differ in an infinite number of places, but which when encoded differ in a finite number of places. This has serious implications for the decoding process because it means that a finite number of channel errors could be translated into an infinite number of decoding errors. This phenomenon is called catastrophic error propagation. Fortunately it is possible to spot catastrophic properties of codes from the existence of common factors in the generator polynomials for the code or in the inability to solve for the code-generating function.

Example of catastrophic error propagation

A convolutional code has generator polynomials

$$g^{(1)}(D) = D + 1$$

$$g^{(0)}(D) = D^2 + 1$$

which have a common factor $D + 1$. An input sequence 1 1 1 1... results in an output sequence 11 01 00 00..., from which we can see that a simple error sequence of finite length could cause reception of 00 00 00 00... which decodes to an all-zero sequence. Thus a finite reception error results in an infinite number of decoding errors, i.e. there is catastrophic error propagation.

2.13   PUNCTURED CONVOLUTIONAL CODES

Although it is possible to define good generator sets for convolutional codes of any rate and to use Viterbi decoding, the computation may become rather complex at high rates. Take, for example a rate 3/4 code for which $v = 3$ and $K = 6$. There are 64 path computations in the decoding of every 4-bit output frame, each involving a 4-bit comparison. Now suppose instead we use a rate 1/2 code and delete two output bits every three frames. For this code, we keep $v = 3$ for comparability, but $K = 4$ giving 16 path computations per frame with three frames required to give the equivalent of the 4-bit output frame of the original. Thus we have 48 comparisons, each of only 1 or 2 bits. Similar considerations show that the computational gains are even greater at higher rates, and are even worthwhile for rate 2/3.

Of course, computational considerations would be worthless if the codes produced by the above process, known as puncturing, did not produce codes of comparable performance. Fortunately, however, there are many punctured codes with a performance which, in terms of coding gain, comes within 0.1 or 0.2 dB of the optimum code. There is therefore little point in using codes other than punctured codes for higher rates. Table 2.15, based on data from [1], shows rate 1/2 codes that can be punctured to produce good rate 2/3 or 3/4 codes. For the rate 2/3 codes, the first two generators (octal) are used to produce the first 2-bit output frame and in the next frame only the third generator (which is the same as one of the other two) is used. For rate 3/4 codes there is then a third frame in which the fourth generator is used.

Apart from the computational consideration, punctured convolutional codes are important because of the possibility of providing several codes of different rates with only one decoder. It would be possible, for example, to operate with a rate 3/4 code in reasonably good reception conditions but allow the transmitter and receiver to agree criteria for a switch to rate 2/3 or 1/2 if noise levels increased and higher values of d_∞ are required. Such a scheme would be known as *adaptive coding*.

Table 2.15   Rate 1/2 codes punctured to rates 2/3 and 3/4

	Generators	d_∞	Generators	d_∞	Generators	d_∞
v	$R = 1/2$		$R = 2/3$		$R = 3/4$	
2	7, 5	5	7, 5, 7	3	7, 5, 5, 7	3
3	15, 17	6	15, 17, 15	4	15, 17, 15, 17	4
4	31, 33	7	31, 33, 31	5	31, 33, 31, 31	3
4	37, 25	6	37, 25, 37	4	37, 25, 37, 37	4
5	57, 65	8	57, 65, 57	6	65, 57, 57, 65	4
6	133, 171	10	133, 171, 133	6	133, 171, 133, 171	5
6	135, 147	10	135, 147, 147	6	135, 147, 147, 147	6
7	237, 345	10	237, 345, 237	7	237, 345, 237, 345	6

2.14   APPLICATIONS OF CONVOLUTIONAL CODES

Convolutional codes have been widely applied to satellite communications. Provided the earth station is well within the antenna footprint, the noise characteristic of communications to geostationary satellites is reasonably represented by the AWGN model, although there are variations with time depending on atmospheric conditions. The main application is digital speech and the required BER is 10^{-5}. All this makes it suitable for convolutional codes and the rate $1/2$ $K = 7$ code described in Section 2.12 has been regularly adopted. The performance of this code was plotted in Chapter 1 (Section 1.9). More recent challenges for satellite communications, however, include the need for nongeostationary orbits to support mobile networks. The codes used for cellular mobile communications will therefore be of interest for satellite communications too.

In cellular mobile communications, the channel characteristic is less favourable with burst errors arising from multipath (reflections), shadowing of the signal and cochannel interference (reuse of the same frequency in other cells), but the need to achieve coding gain at moderate target bit error rates again dictates that convolutional codes should be used. Because of the hostile channel environment, the voice coders (vocoders) are designed to work well with bit error rates of 10^{-3} and acceptably with error rates well above this.

The GSM standard for digital mobile communications is a time division multiple access (TDMA) system providing a bit rate on each channel of 22 800 bits per second. This is achieved in time-slots which hold 114 bits of data (in fact the term *burst* is used for the time-slots, but as mobile channels suffer bursty errors this would result in two possible meanings of the word). The principal application is digital voice with vocoders that can produce acceptable quality even in the presence of bit errors at a rate of 1% or more. To deliver coding gain at this level, convolutional codes are needed with interleaving to protect against the channel error bursts. The code used is rate $1/2$ $K = 5$ with generators

$$g^{(1)}(D) = D^4 + D^3 + 1$$

$$g^{(0)}(D) = D^4 + D^3 + D + 1$$

The original full rate (FR) voice coding standard for GSM used a 13 000 bit per second vocoder operating with 20 mS frames, i.e. 260 bits per frame. Vocoded speech consists partly of filter parameters and partly of excitation parameters to generate the speech at the receiver. The subjective effects of errors depend on the parameters affected and the bits were accordingly classified into 182 class 1 bits (sensitive) and 78 class 2 bits (not sensitive). Of the class 1 bits, 50 (known as class 1a) were considered to be the most important and able to be predicted from past values. They were protected by a 3-bit CRC to allow for some error detection and error concealment after decoding. The class 1a bits, the 3-bit CRC and the remaining class 1 bits (class 1 b) were then fed into the convolutional encoder, followed by four zeros acting as flushing bits to clear the encoder memory. The encoder produces 378 bits $[2 \times (178 + 3 + 4)]$ and the class 2 bits, uncoded, make this up to 456 bits.

To protect against burst errors a scheme known as block diagonal interleaving is used. It incorporates an element of convolutional interleaving in which the odd bits are delayed by four blocks before an interleave pattern which maintains the separation into even- and odd-numbered bits. The even-numbered bits of the eight blocks are interleaved into the even-numbered bits of eight time-slots and odd-numbered bits into odd-numbered bits of eight time-slots, but starting four time-slots later.

The FR standard vocoder is replaced in newer terminals by the EFR standard which produces higher quality at a slightly lower bit rate of 12 200 bits per second. There are therefore 244 bits in each frame and an extra 16 bits are created by a preliminary channel coding stage to give extra error protection. The preliminary coding creates an 8-bit CRC on the 65 most important bits (the 50 class 1a bits and 15 of the class 1b bits) and also puts a (3, 1) binary repetition code on each of four class 2 bits judged to be the most important.

The UMTS, IS-95 and CDMA2000 standards for mobile communications use a $K = 9$ convolutional code. For rate $1/2$ the generator polynomials are

$$g^{(0)}(D) = D^8 + D^4 + D^3 + D^2 + 1$$

$$g^{(1)}(D) = D^8 + D^7 + D^5 + D^3 + D^2 + D + 1$$

and for rate $1/3$ they are

$$g^{(0)}(D) = D^8 + D^7 + D^6 + D^5 + D^3 + D^2 + 1$$

$$g^{(1)}(D) = D^8 + D^7 + D^4 + D^3 + D + 1$$

$$g^{(2)}(D) = D^8 + D^5 + D^2 + D + 1$$

These codes are likely to find their way into future satellites too.

2.15   CODES FOR MULTILEVEL MODULATIONS

For multilevel modulations, the choice of a good convolutional code has to be judged, not by its free distance, but by the minimum value of squared Euclidean distance between code sequences when the code is used in conjunction with the modulation. The initial design of codes of this type was undertaken by Ungerboeck [2] who adopted a set-partitioning approach to the mapping of symbol values on the constellation. Consider, for example, an 8-PSK constellation as shown in Figure 2.10. The leftmost bit corresponds to the principal set partition, with adjacent points separated into different sets. The next bit then partitions closest points in each set into different subsets. The rightmost bit denotes the point within the subset.

Ungerboeck's approach was to use a specially designed convolutional code to protect the set partition and leave the remaining bit(s) uncoded. In the 8-PSK example, a rate $1/2$ code would be used for the set partition with one bit left uncoded, so that the overall code rate is $2/3$ and the throughput equivalent to uncoded QPSK.

The square of the distance between the points corresponding to a change in the uncoded bit, however, is twice that of the distance between points in the QPSK constellation, so that, at best, a 3 dB coding gain is achievable if the code protecting the set partition is sufficiently strong. In practice it is easy to find a code that will give this level of performance.

This approach can be generalized to higher order constellations and more stages of partitioning. For example, with 4 or more bits per symbol, it would be possible to partition three times and to use a rate 2/3 code to protect the partition. The limits to what can be achieved are shown in Table 2.16, again expressed relative to the performance of uncoded BPSK/QPSK. It should be noted that high gains from high rate codes may be difficult to achieve in practice.

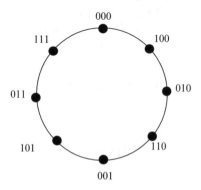

Figure 2.10   Set-partitioned 8-PSK constellation

Table 2.16   Upper Bounds on performance of Ungerboeck-coded MPSK

m	2 partitions rate 1/2 (dB)	3 partitions rate 2/3 (dB)	4 partitions rate 3/4 (dB)
1			
2			
3	3.01		
4	1.76	4.77	
5	-2.32	3.01	6.02
6	-7.21	-1.35	3.98
7	-12.39	-6.41	-0.56
8	-17.73	-11.72	-5.74

Ungerboeck-coded 8-PSK

Consider the 8-PSK constellation with set partitioning shown in Figure 2.11. The squared Euclidean distances between points are shown with the distance between the closest points of a QPSK constellation set to 1, so that the performance of rate 2/3 coded 8-PSK can easily be compared with uncoded QPSK.

A simple convolutional encoder, with effective rate 2/3, often used in conjunction with this constellation is shown in Figure 2.12. The rightmost bit is uncoded and a rate 1/2 code used to determine the other two bit values.

The state diagram for this code is shown in Figure 2.13. Note that there are two paths between each pair of states, depending on the value of the uncoded bit. The squared Euclidean distance to each corresponding transmitted point is shown in parentheses.

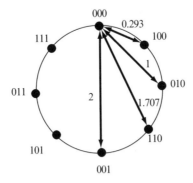

Figure 2.11 Squared Euclidean distances in 8-PSK constellation

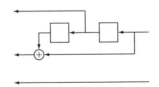

Figure 2.12 Encoder for 8-PSK

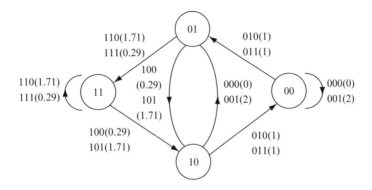

Figure 2.13 State diagram (with squared Euclidean distances)

The minimum weight loop leaving and returning to state 00, with weight interpreted as squared Euclidean distance, is the loop 00–01–10–00 with minimum squared Euclidean distances of 1, 0.29 and 1, respectively, for the three transitions. The squared Euclidean distance to the closest code path is therefore 2.29, corresponding to a coding gain of 3.6 dB, if the uncoded bit is ignored. However the loop on state 00 resulting from a nonzero uncoded bit is at a squared Euclidean distance of 2 from the all-zero loop, so that the uncoded bits are less well protected than the encoded ones and the overall coding gain is 3.0 dB. This indicates the possibility that better codes could be found which give some protection to all the bits.

Although the approach of leaving some bits uncoded is widely adopted, codes are now known which allow all the bits in the constellation to be encoded and yielding improved gains. For example the rate 2/3 encoder with the polynomials

$$g_1^{(2)}(D) = D, \; g_1^{(1)}(D) = D^2, \; g_1^{(0)}(D) = 1$$

$$g_0^{(2)}(D) = 0, \; g_0^{(1)}(D) = 1, \; g_0^{(0)}(D) = D$$

has a free squared Euclidean distance of 2.29 corresponding to a coding gain of 3.6 dB. The asymptotic coding gains available for various rate 2/3 codes on 8-PSK, relative to uncoded QPSK are shown in Table 2.17.

Table 2.17   Coded 8-PSK asymptotic performance

No. of states	No. of bits encoded	Gain (dB)
4	1	3.01
8	2	3.6
16	2	4.13
32	2	4.59
64	2	5.01
128	2	5.17
256	2	5.75

QAM constellations

Another way to expand the signal constellation without bandwidth penalties is to adopt Quadrature Amplitude Modulation (QAM). Two signals in phase quadrature are combined such that the resultant can take a large number of discrete amplitude and phase combinations, as shown in Figure 2.14 for a 16-level constellation. The number of points in the constellation may be doubled with approximately the same mean symbol energy by insertion of a number of intermediate points. The distance between the points is now reduced by a factor of $\sqrt{2}$, so that symbol energy must be doubled for the same symbol error rate. In fact to maintain E_b/N_0 the symbol energy will be increased by a factor of $[1 + \log_2(M)]/\log_2(M)$. In the limit of large M, this increase becomes negligible and the reduction of distance between the closest points associated with each doubling of M produces a 3 dB loss.

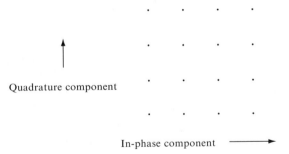

Figure 2.14   16-QAM constellation points

0000	1000	0010	1010
1100	0100	1110	0110
0011	1011	0001	1001
1111	0111	1101	0101

Figure 2.15   Partitioning of 16-QAM

The partitioning of QAM constellations follows a similar approach except that care needs to be taken in identifying the closest points within the constellation. For example, a 16-point constellation might be mapped as in Figure 2.15. In the first partition the closest points are separated horizontally and vertically, but for the next partition the closest points within a single set are separated diagonally.

Higher order QAM constellations often have the distribution of points slightly rearranged from the square layout shown. The purpose of the adjustment is to reduce the average signal energy, or the variance of the signal energy, for a given spacing of points. Other multilevel constellations, such as pairs of PSK rings one inside the other, are also found.

The classic application of Ungerboeck codes is for modems communicating over telephone lines. The bandwidth is limited to less than 4 kHz, making the use of trellis-coded modulation necessary to achieve the data rates that are theoretically possible. Because linearity is not a problem with low power transmissions, QAM constellations are used. Data rates up to 33.6 kb/s are achievable in this way using the V.34 standard. The higher data rates (up to 56 kb/s) from digital modems are achieved by treating the data as if it was 7-bit PCM samples from speech sampled at 8 kHz. This is used in the asymmetric V.90 modem where a V.34 analogue standard is used for the uplink and a digital modem for the downlink.

2.16   SEQUENTIAL DECODING

The Viterbi algorithm is satisfactory only for codes of relatively short constraint lengths because of the number of paths that must be updated and stored. If we examined the actions of a Viterbi decoder, we would usually find that a small number of paths are established as being the most likely with the other stored paths being much less likely. Sequential decoding aims to simplify the decoding task by concentrating the search on the most likely paths. In this way it can use codes with much longer constraint lengths. There are several slightly different implementations of sequential decoding, but two of them – the Fano algorithm and the stack algorithm – are illustrative of the basic approaches that are possible.

The Fano algorithm works frame-by-frame, examining the received sequence, deciding the most likely code frame and advancing to the appropriate point in the trellis. The metric adopted for each received bit is

$$\gamma_{ij} = \log_2 \left[\frac{p(r_{ij} \mid c_{ij})}{p(r_{ij})} \right] - R$$

The branch metric is

$$\gamma_j = \sum_{i=1}^{n_0} \gamma_{ij}$$

and the path metric is

$$\Gamma(l) = \sum_{j=1}^{l} \gamma_j$$

where j represents the frame number, i the bit within the frame, r_{ij} the received bit, c_{ij} the appropriate bit of the code path being followed and R the code rate. This quantity is called the Fano metric.

Note that the Fano metric depends on the noise level of the channel. Assuming hard decisions with a channel bit error rate of 10^{-2}, the value of the Fano metric associated with a bit correctly received would be 0.471 and for an incorrect bit -5.144; these may be scaled and approximated to $+1$ and -11. Thus when correct values are being received, the metric is increasing slowly; however, when an error occurs it decreases sharply. Following the wrong path in the trellis results in poor correlation between received bits and possible trellis paths, so that the metric will decrease.

At each stage in decoding, the path metric is compared with a running threshold to decide whether the decoder should move forwards or backwards. On the first visit to any node in the trellis, the threshold is set as tightly as possible. Generally the metric should be no greater than the threshold and if the threshold is exceeded, the decoder has to decide whether to backtrack or to loosen the threshold. The latter action is taken when backtracking does not bring the metric down below the current threshold. Changes to the threshold are made in multiples of some fixed value Δ which is a design parameter. Too small a value of Δ will increase the amount of backtracking, too large and the decoder will follow incorrect paths for a considerable distance before backtracking begins. A flowchart of the Fano algorithm is shown in Figure 2.16.

The stack algorithm eliminates the tendency of the Fano algorithm to visit certain nodes several times, thus reducing computation, but at the expense of increased storage. A number of previously examined paths and their accumulated metrics are held on a stack, the path with the lowest metric at the top. The decoder takes the path at the top, creates 2^{k_0} successor paths, computes their metrics and places them in the appropriate positions on the stacks. The stack may overflow, but the paths lost will have high metrics and are unlikely to figure in the maximum likelihood solution. The reordering of paths is also a problem, but Jelinek [3] proposed an approach which has been widely adopted.

Both the Fano and the stack methods operate within a fixed decoding window, as do Viterbi decoders, and output the first frame from a full decoding window. There are other distinct characteristics of sequential decoding which may affect its suitability for particular applications. The speed of the decoding will depend on the method used, but there is variability too in decoding speed which can cause problems for real time operation. This variability may be smoothed out by provision of a buffer for the incoming frames, but in certain circumstances it may be necessary to force the decoder along a particular path or to abandon the decoding of a section of the code. This is not necessarily a bad thing; correct design can ensure that problems of this type occur mainly when the error rates are so high that decoding errors are in any case highly likely, and at least it is known that a decoding problem has occurred. Nevertheless, sequential decoding is more at home with non-real time and off-line applications.

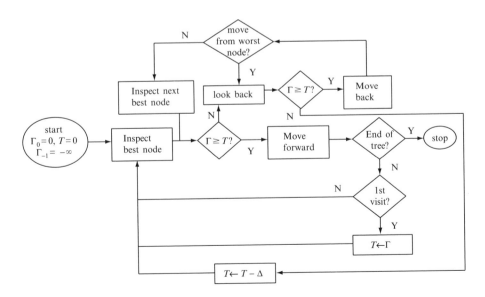

Figure 2.16 Fano algorithm

The principal advantage of sequential decoding is that the reduced computation allows for the use of greater constraint lengths, values around 40 being commonly quoted. This means that more powerful codes with higher values of free distance and higher coding gains can be employed. Soft-decision decoding is simple to incorporate in principle and has advantages in terms of reduced backtracking. On the other hand, it does require the increased cost of a soft-decision demodulator and the increased coding gain may be more easily obtained by increasing the constraint length. As a result it is less common to find soft decisions with sequential decoding than is the case with Viterbi decoding.

2.17 CONCLUSION

Convolutional codes are extensively treated in most standard books on error control coding [4–8]. A detailed text on convolutional codes is also available [9]. Further developments in the coding of multilevel modulations include the use of punctured codes to achieve the appropriate rate and the Euclidean distance properties for the modulation [10] and the design of codes to work with Gray-coded modulations. For Rayleigh fading channels it is Hamming distance, not Euclidean distance that is important and so ordinary convolutional codes can be mapped onto the Gray-coded modulation through an interleaver to combat error bursts [11].

Recursive systematic convolutional codes, used with iterative decoding and commonly known as turbo codes, are treated in Chapter 10.

2.18 EXERCISES

1 An encoder has the following generator polynomials:

$$g_1^{(2)}(D) = D, \; g_1^{(1)}(D) = D^2, \; g_1^{(0)}(D) = 1$$
$$g_0^{(2)}(D) = 0, \; g_0^{(1)}(D) = 1, \; g_0^{(0)}(D) = D$$

Draw the encoder schematic diagram. Encode the information sequence 10 01 00 01.

2 For the code of question 1, quantify the following terms:
   (a) input frame
   (b) output frame
   (c) input constraint length
   (d) output constraint length
   (e) memory order
   (f) memory constraint length

3   An encoder has generator polynomials

$$g^{(1)}(D) = D^3 + D^2 + 1$$

$$g^{(0)}(D) = D^3 + D^2 + D + 1$$

Draw the encoder schematic and the state diagram. Find the value of d_∞.

4   For the code of question 3, find the lengths and input weights of all paths with the three lowest values of output weight. Hence find the corresponding terms of the generating function for the code.

5   A code has generators

$$g^{(1)}(D) = D^2 + 1$$

$$g^{(0)}(D) = D^2 + D$$

From the state diagram, or by considering the encoding of the sequence 1111..., deduce that the code exhibits catastrophic error propagation.

6   From the encoder state diagram below, find the generator polynomials of the code.

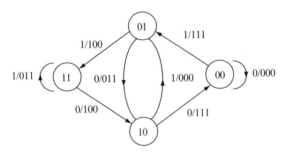

7   For the convolutional encoder of Figure 2.1, decode the hard-decision sequence 11 01 11 00 10 00 01 11 11.

8   Find the hard-decision coding gains at bit error rates of 10^{-3} and 10^{-5} for the rate 1/2 $K = 3$ code whose weight structure is analysed in Section 2.7. How much does the gain change if unquantized soft-decision decoding is used?

9   Repeat question 6 for the rate 1/2 $K = 7$ code described in Section 2.12.

10 Prove that for a rate R convolutional code operating on a binary PSK channel (bit error rate given in Equation (1.1)), the coding gain at low bit error rates is of the order of $Rd_\infty/2$ for hard-decision decoding and Rd_∞ for soft-decision decoding.

11   Devise a partitioning scheme for 16-ary PSK.

12   An uncoded communication channel uses 8-ary modulation. It is decided to go over to an Ungerboeck-coded 16-ary channel with two-stage partitioning. Compare the expected coding gains at constant symbol error rate if the constellation is MPSK against those for QAM. What would be the effects of three-stage partitioning?

13   Verify the value of free squared Euclidean distance for the rate 2/3 code with polynomials

$$g_1^{(2)}(D) = D, \; g_1^{(1)}(D) = D^2, \; g_1^{(0)}(D) = 1$$

$$g_0^{(2)}(D) = 0, \; g_0^{(1)}(D) = 1, \; g_0^{(0)}(D) = D$$

used on set partitioned 8-PSK.

2.19   REFERENCES

1   J.B. Cain, G.C. Clark and J.M. Geist, *Punctured convolutional codes of rate $(n-1)/n$ and simplified maximum likelihood decoding*, IEEE Transactions on Information Theory, Vol. IT-25, pp. 97–100.

2   G. Ungerboeck, *Channel coding with multilevel/phase signal*, IEEE Transactions on Information Theory, Vol. 28, pp. 55–66, January 1982.

3   F. Jelinek, *A Fast Sequential Decoding Algorithm Using a Stack*, IBM Journal of Research and Development, Vol. 13, pp. 675–685, 1969.

4   S. Lin and D.J. Costello, *Error Control Coding: Fundamentals and Applications*, Prentice Hall, 1983.

5   G.C. Clark and J.B. Cain, *Error-Correction Coding for Digital Communications*, Plenum Press, 1981.

6   A.M. Michelson and A.H. Levesque, *Error-Control Techniques for Digital Communication*, John Wiley & Sons, 1985.

7   S.B. Wicker, *Error Control Systems for Digital Communication and Storage*, Prentice Hall, 1994.

8   M. Bossert, *Channel Coding for Telecommunications*, John Wiley & Sons, 1999.

9   L.H.C. Lee, *Convolutional Coding: fundamentals and applications*, Artech House, 1997.

10   E. Zehavi, '8-PSK trellis codes for a Rayleigh fading channel', IEEE Transactions on Communications, Vol. 40, pp. 873–883, May 1992.

11   L.H.C Lee, 'New rate-compatible punctured convolutional codes for Viterbi decoding', IEEE Transactions on Communications, Vol. 42, pp. 3073–3079, December 1994.

3

Linear block codes

3.1 INTRODUCTION

This chapter presents the most important aspects of linear block codes. Block codes were defined in Section 1.4 and an example of a block code was used extensively in the first chapter. Almost all useful block codes possess the property of linearity, which was defined in Section 1.4.2. The topics to be covered in this chapter include the ways in which codes can be defined, the uses of linearity in the encoding and (hard-decision) decoding operations, minimum distance and the bounds on distance that apply to block codes. It is largely assumed that forward error correction is required because the detection of errors is achieved as the first step of error correction and thus is included in the techniques described.

It is assumed that the reader is familiar with the material of Sections 1.4 and 1.8. The relationship between minimum distance and error detection and correction properties of block codes was covered in Section 1.14.3. Only binary codes are treated in this chapter, enabling the mathematics of nonbinary codes to be left until Chapter 5. The special techniques associated with the subset of linear block codes known as cyclic codes will be covered in Chapter 4.

3.2 MATHEMATICS OF BINARY CODES

The mathematics of coding can be rather complicated if all classes of codes are to be studied. By restricting ourselves, for the moment, to simple codes we can employ simple mathematics to gain familiarity with the subject before attempting the more difficult codes. As a result this chapter will require nothing more difficult than an understanding of matrix representation of equations and the application of simple logical functions.

The main reason that the mathematics of coding can appear complicated is that we need to be able to carry out arithmetic in what is called a *finite field*. Any code consists of a number of symbols which can take only certain values, the simplest example of a symbol being a bit which can take only two values although other symbols with more levels can be devised. It is necessary to define our arithmetic operations in a way that only valid symbol values can be produced. A finite field is a defined set of values plus two defined operations and their inverses which can yield only values within the set.

The operations to be carried out to produce linear codes defined over a finite set of values are called addition and multiplication, and their inverses may be thought of as subtraction and division. The operations themselves will not, however, correspond to our normal understanding of those terms. For all codes, the definition of the appropriate arithmetic is necessary before the encoding and decoding can be explained. For nonbinary codes these definitions are not straightforward.

Fortunately there is only one important family of nonbinary codes, namely the Reed Solomon codes, although for a proper understanding of some other codes a finite field approach is valuable. Nevertheless, we can go a long way dealing only with binary fields, for which the appropriate arithmetic is simply modulo-2:

$$0 + 0 = 0$$
$$0 + 1 = 1$$
$$1 + 1 = 0$$

$$0 \times 0 = 0$$
$$0 \times 1 = 0$$
$$1 \times 1 = 1$$

The inverse of addition (subtraction) is equivalent to addition, division by zero is not allowed and division by 1 is equivalent to multiplication by 1. Our finite field arithmetic will therefore be rather easy, with the only matter of note being the modulo-2 addition corresponding to the exclusive-OR function of Boolean logic.

3.3   PARITY CHECKS

To obtain an insight into how a linear code might be produced, let us take a simple example in which a codeword is produced from the information by letting the information flow directly through into the codeword and then following it with a single bit calculated from all the information bits. We shall consider two methods of calculating this final bit:

(i) The final bit is set such that the modulo-2 sum of all the bits in the codeword is 1.

(ii) The final bit is set such that the modulo-2 sum of all the bits in the codeword is 0.

In the first case the codeword is said to have odd parity, i.e. there is an odd number of ones in the codeword. In the second case there is an even number of ones in the codeword which therefore has even parity. The extra bit is called a parity check bit and may be called an odd parity or even parity check as appropriate.

The odd and even parity codes are shown in Tables 3.1 and 3.2, respectively, for the case in which there are three information bits.

Table 3.1   Odd parity code

Information	Code
0 0 0	0 0 0 1
0 0 1	0 0 1 0
0 1 0	0 1 0 0
0 1 1	0 1 1 1
1 0 0	1 0 0 0
1 0 1	1 0 1 1
1 1 0	1 1 0 1
1 1 1	1 1 1 0

Table 3.2   Even parity code

Information	Code
0 0 0	0 0 0 0
0 0 1	0 0 1 1
0 1 0	0 1 0 1
0 1 1	0 1 1 0
1 0 0	1 0 0 1
1 0 1	1 0 1 0
1 1 0	1 1 0 0
1 1 1	1 1 1 1

We note that the code of Table 3.1 does not contain the all-zero sequence which must be part of a linear code. Thus the odd parity check produces a nonlinear code. On the other hand, the code of Table 3.2 is linear; systems which produce even parity checks on some or all of the bits result in a linear code. Note that in this case the parity check bit is the modulo-2 sum of the bits from which it is calculated. Thus the parity for the information sequence 101 is the modulo-2 sum of 1, 0 and 1, i.e. 0.

3.4   SYSTEMATIC CODES

The above examples have had the property that the information bits appear in the codeword unchanged with some parity bits added. A particularly common arrangement is that the information appears at the start of the codeword and is followed by the parity check bits. In this case the code is said to be systematic. Any linear block code can be put into systematic form and at worst is only trivially different from a systematic arrangement in that a fixed change to the order of the symbols can produce the systematic form. A linear block code can therefore always be considered as *equivalent* to a systematic code.

3.5   MINIMUM HAMMING DISTANCE OF A LINEAR BLOCK CODE

It was stated in Chapter 1 that a consequence of linearity is that the distance structure of the code appears the same regardless of which codeword it is viewed from. If **u**, **v** and **w** are codewords and $\mathbf{d}(\mathbf{u}, \mathbf{v})$ signifies the distance between **u** and **v**, then

$$\mathbf{d}(\mathbf{u}, \mathbf{v}) = \mathbf{d}(\mathbf{u} + \mathbf{w}, \mathbf{v} + \mathbf{w}) \qquad (3.1)$$

The sequences **u** + **w** and **v** + **w** are codewords and so the relationship between **u** and **v** is repeated at other points in the code. In particular we can set **w** = **v** to give

$$\mathbf{d}(\mathbf{u}, \mathbf{v}) = \mathbf{d}(\mathbf{u} + \mathbf{v}, \mathbf{0}) \qquad (3.2)$$

Thus we can say that the distance between any pair of codewords is the same as the distance between some codeword and the all-zero sequence. We can therefore reach the following conclusion:

The minimum distance of a linear block code is equal to the minimum number of nonzero symbols occurring in any codeword (excluding the all-zero codeword).

The number of nonzero symbols in a sequence is called the weight of the sequence, and so the minimum distance of a linear block code is equal to the weight of the minimum weight codeword.

3.6   HOW TO ENCODE – GENERATOR MATRIX

In the previous examples of codes, we have used a table to hold all the codewords and looked up the appropriate codeword for the required information sequence. We can, however, create codewords by addition of other codewords, which means that there is no need to hold every codeword in a table. If there are k bits of information, all we need is to hold k linearly independent codewords, i.e. a set of k codewords none of which can be produced by linear combinations of two or more codewords in the set. The easiest way to find k linearly independent codewords is to choose those which have 1 in just one of the first k positions and 0 in the other $k - 1$ of the first k positions. Using, for instance, the example (7, 4) code of chapter 1, we need just the four codewords below:

$$
\begin{array}{ccccccc}
1 & 0 & 0 & 0 & 1 & 1 & 0 \\
0 & 1 & 0 & 0 & 1 & 0 & 1 \\
0 & 0 & 1 & 0 & 0 & 1 & 1 \\
0 & 0 & 0 & 1 & 1 & 1 & 1 \\
\end{array}
$$

If we wish, for example, to obtain the codeword for 1011, we add together the first, third and fourth codewords in the list to give 1011010.

The process of encoding by addition can be represented in matrix form by

$$\mathbf{v} = \mathbf{u} \, \mathbf{G} \tag{3.3}$$

where **u** is the information block, **v** the codeword and **G** the generator matrix.

Taking the above example, we can represent the code by the generator matrix below:

$$\mathbf{G} = \begin{bmatrix} 1 & 0 & 0 & 0 & 1 & 1 & 0 \\ 0 & 1 & 0 & 0 & 1 & 0 & 1 \\ 0 & 0 & 1 & 0 & 0 & 1 & 1 \\ 0 & 0 & 0 & 1 & 1 & 1 & 1 \end{bmatrix}$$

If, as before, we wish to encode the sequence 1 0 1 1 we obtain

$$\mathbf{v} = \begin{bmatrix} 1 & 0 & 1 & 1 \end{bmatrix} \begin{bmatrix} 1 & 0 & 0 & 0 & 1 & 1 & 0 \\ 0 & 1 & 0 & 0 & 1 & 0 & 1 \\ 0 & 0 & 1 & 0 & 0 & 1 & 1 \\ 0 & 0 & 0 & 1 & 1 & 1 & 1 \end{bmatrix}$$

$$\mathbf{v} = \begin{bmatrix} \mathbf{1} & \mathbf{0} & \mathbf{1} & \mathbf{1} & \mathbf{0} & \mathbf{1} & \mathbf{0} \end{bmatrix}$$

Note that the generator is a $k \times n$ matrix where k is the dimension of the code (number of information bits) and n is the length of any codeword. In this case the generator has a special form corresponding to a systematic code. It consists of a $k \times k$ unit matrix followed by a $k \times (n - k)$ matrix of parity check bits.

If we were going to use the generator matrix approach to encoding of a systematic code, there would be no point in storing that part of the codewords that corresponds to the information. We therefore need only to store $k \times (n - k)$ bits in some form of read-only memory (ROM), and let the information bits determine which of the $(n - k)$-bit sequences are to be modulo-2 added to form the parity checks of the codeword.

3.7 ENCODING WITH THE PARITY CHECK MATRIX

In Section 3.3 we introduced the idea of a parity check and deduced that even parity checks corresponded to a linear encoding operation. It should therefore be possible to define a code in terms of groups of bits which must be of even parity, i.e. their modulo-2 sum must be zero. For example, we may choose to calculate three parity check bits from four information bits as shown below. The leftmost bit is considered to be bit 6 and the rightmost bit 0, so that the information corresponds to bits 6 to 3 and the parity checks are bits 2 to 0:

$$\text{bit } 2 = \text{bit } 6 \oplus \text{bit } 5 \oplus \text{bit } 3$$
$$\text{bit } 1 = \text{bit } 6 \oplus \text{bit } 4 \oplus \text{bit } 3$$
$$\text{bit } 0 = \text{bit } 5 \oplus \text{bit } 4 \oplus \text{bit } 3$$

In other words, bits 6, 5, 3 and 2 form an even parity group, as do bits 6, 4, 3 and 1 and bits 5, 4, 3 and 0. If the information is 1011 then

$$\text{bit } 6 = 1 \quad \text{bit } 5 = 0 \quad \text{bit } 4 = 1 \quad \text{bit } 3 = 1$$

from which we can calculate

$$\text{bit } 2 = 0 \quad \text{bit } 1 = 1 \quad \text{bit } 0 = 0$$

The codeword is therefore 1011010, as was the case for the example in Section 3.6. A check of the codewords forming the rows of the generator matrix in that section will confirm that this system of parity checks in fact generates the same code. The way in which a code specified by a generator matrix can be transformed to an equivalent system of parity checks will shortly become apparent.

The system of parity checks can be put into the matrix representation below:

$$\mathbf{H} = \begin{bmatrix} 1 & 1 & 0 & 1 & 1 & 0 & 0 \\ 1 & 0 & 1 & 1 & 0 & 1 & 0 \\ 0 & 1 & 1 & 1 & 0 & 0 & 1 \end{bmatrix}$$

The matrix $\mathbf{H}$ is called the parity check matrix and each row represents an even parity group with ones in the positions of the bits that comprise the group.

Because the rows of the parity check matrix correspond to even parity groups, the scalar product of any codeword with any row will be zero. The generator matrix has rows which are themselves codewords. Thus if we form the scalar product of any row of the generator matrix with any row of the parity check matrix the result will be zero. Matrix multiplication is carried out, however, by forming scalar products of the rows of the first matrix with columns of the second. We can therefore write

$$\mathbf{G}\,\mathbf{H}^T = \mathbf{0} \tag{3.4}$$

We can now see how to form the parity check matrix and thus how to formulate a code in terms of parity checks. It is a $(n - k) \times n$ matrix constructed in such a way that Equation (3.4) is satisfied. Starting from the generator matrix, separate the $k \times (n - k)$ section corresponding to the parity checks:

$$\begin{bmatrix} 1 & 1 & 0 \\ 1 & 0 & 1 \\ 0 & 1 & 1 \\ 1 & 1 & 1 \end{bmatrix}$$

transpose it

$$\begin{bmatrix} 1 & 1 & 0 & 1 \\ 1 & 0 & 1 & 1 \\ 0 & 1 & 1 & 1 \end{bmatrix}$$

and follow it with a $(n - k) \times (n - k)$ unit matrix

$$\mathbf{H} = \begin{bmatrix} 1 & 1 & 0 & 1 & 1 & 0 & 0 \\ 1 & 0 & 1 & 1 & 0 & 1 & 0 \\ 0 & 1 & 1 & 1 & 0 & 0 & 1 \end{bmatrix}$$

This form again assumes a systematic code.

This particular parity check matrix has one further feature of note. Looking at the columns of the matrix we see that all possible 3-bit patterns of 1s and zeros are to be found, with the exception of the all-zero pattern. This feature is characteristic of the family of codes to which this code belongs, namely the Hamming codes. We can construct Hamming codes with any number of parity check bits by making a matrix with $n - k$ rows and with the columns consisting of all the $2^{n-k} - 1$ possible patterns of $n - k$ bits which exclude the all-zero pattern. For example the parity check matrix

$$\mathbf{H} = \begin{bmatrix} 1 & 1 & 0 & 1 & 0 & 0 & 1 & 1 & 1 & 0 & 1 & 1 & 0 & 0 & 0 \\ 1 & 0 & 1 & 0 & 1 & 0 & 1 & 1 & 0 & 1 & 1 & 0 & 1 & 0 & 0 \\ 0 & 1 & 1 & 0 & 0 & 1 & 1 & 0 & 1 & 1 & 1 & 0 & 0 & 1 & 0 \\ 0 & 0 & 0 & 1 & 1 & 1 & 0 & 1 & 1 & 1 & 1 & 0 & 0 & 0 & 1 \end{bmatrix}$$

defines a (15, 11) Hamming code. The order of the columns is immaterial as far as the definition of a Hamming code is concerned, although we may wish to preserve the unit matrix on the right corresponding to the systematic form.

The form of the parity check matrix gives the Hamming code some special decoding properties that will be seen in the next section.

3.8 DECODING WITH THE PARITY CHECK MATRIX

Decoding generally consists of two stages. The first is to check whether the sequence corresponds to a codeword. If only error detection is required, then this completes the decoding process. If error correction is required, then there must be an attempt to identify the error pattern. This second stage is likely to be much more complex than the first and its implementation will normally be the major factor in the overall complexity, speed and cost of the encoder/decoder (codec).

Error detection involves deciding whether all the even parity checks are satisfied in the received sequence. If we perform modulo-2 addition on all the even parity groups, the result will be zero for those that are satisfied and one for those that are not. The resulting $(n - k)$-bit result is called the *syndrome*. An alternative, but equivalent definition of syndrome is that it is the sequence formed by modulo-2 adding the received parity bits to the parity bits recalculated from the received information. If the received sequence is $\mathbf{v}'$ then the syndrome can also be regarded as a vector $\mathbf{s}$ where

$$\mathbf{s} = \mathbf{v}' \mathbf{H}^T \tag{3.5}$$

An all-zero syndrome indicates that the sequence is correct. Any other syndrome indicates the presence of errors. Because of the linear properties of the code, any received sequence can be considered to be the sum of a codeword and an error pattern and the syndrome is likewise the sum of that for the codeword (i.e. zero) and that for the error pattern. This leads to the result that *the syndrome value depends only on the errors, not on the transmitted codeword.*

The equivalence of the above three definitions of syndrome can be verified by taking an example. Suppose we receive a sequence 1000101, calculating the syndrome by each of the three methods gives:

(i)

$$\text{bit } 6 \oplus \text{bit } 5 \oplus \text{bit } 3 \oplus \text{bit } 2 = 0$$
$$\text{bit } 6 \oplus \text{bit } 4 \oplus \text{bit } 3 \oplus \text{bit } 1 = 1$$
$$\text{bit } 5 \oplus \text{bit } 4 \oplus \text{bit } 3 \oplus \text{bit } 0 = 1$$

(ii)

$$\text{received information} = 1000$$
$$\text{recalculated parity} = 110$$
$$\text{received parity} = 101$$
$$\text{syndrome} = 011$$

(iii)

$$\mathbf{s} = \begin{bmatrix} 1 & 0 & 0 & 0 & 1 & 0 & 1 \end{bmatrix} \begin{bmatrix} 1 & 1 & 0 \\ 1 & 0 & 1 \\ 0 & 1 & 1 \\ 1 & 1 & 1 \\ 1 & 0 & 0 \\ 0 & 1 & 0 \\ 0 & 0 & 1 \end{bmatrix} = \begin{bmatrix} 0 & 1 & 1 \end{bmatrix}$$

We need also to be able to relate the syndrome to the errors that have occurred. In the above example, we need to find a bit that is not involved in the first parity check but is involved in the second and third parity checks. If we can find such a bit, we will have identified the position of the error because it will explain the syndrome we have obtained. Looking at the columns of the parity check matrix, we see that bit 4 satisfies the required properties. The columns of the parity check matrix therefore have an important interpretation: the first column is the syndrome of an error in the first bit; likewise any column *m* contains the syndrome of an error in position *m*. Because for a Hamming code the columns contain all the nonzero syndromes, we can relate any syndrome to a single-bit error. Thus if the syndrome is 0 1 1, as in this case, we know that the error is in bit four and we can correct the received sequence 1000101 to 1010101, which is indeed a codeword.

To design a decoder, we could use combinational logic to look for the syndromes corresponding to each position to tell us which bit was in error. Alternatively we could store a number of error patterns ordered according to their syndrome, and merely select the appropriate pattern once the syndrome has been formed.

What happens if two bits are in error? Assuming that bits 3 and 2 were wrong, the syndrome would be 1 1 1 + 1 0 0 = 0 1 1. This would be interpreted in the decoder as an error in bit 4. Because all the syndromes are contained in the parity check matrix, the decoder will always think that it knows what has occurred, even when it is wrong. The Hamming code is good for detecting and correcting single errors per block (minimum distance = 3), but any more errors will always cause a decoding error. Of course we do not expect a decoder to be able to cope with errors beyond half the minimum distance, but the fact that it always fails is a special property of Hamming codes.

3.9   DECODING BY STANDARD ARRAY

Another way to look at the decoding process is to list all the received sequences in sets, each set containing just one codeword that should be the decoder output for any of the sequences in the set. For a binary code there are 2^k codewords and 2^n possible received sequences. The received sequences are therefore partitioned into sets of 2^{n-k}, each set containing one codeword and each sequence in the set having a different syndrome. If the sets are arranged in columns with the codeword at the top, and with all the sequences at a given position in every set having the same syndrome, then the result is an array known as the *standard array*.

Let us consider the example of the (5, 2) code from Chapter 1 (Section 1.8.1). The standard array could be

$$
\begin{array}{cccc}
00000 & 01011 & 10101 & 11110 \\
10000 & 11011 & 00101 & 01110 \\
01000 & 00011 & 11101 & 10110 \\
00100 & 01111 & 10001 & 11010 \\
00010 & 01001 & 10111 & 11100 \\
00001 & 01010 & 10100 & 11111 \\
11000 & 10011 & 01101 & 00110 \\
10010 & 11001 & 00111 & 01100 \\
\end{array}
$$

The way in which this was constructed is as follows. The top row consists of all the codewords. The code has minimum distance of 3, so we expect to detect and correct all single-bit errors, a total of five error patterns. The second row consists, therefore of all the patterns with an error in the first bit, the third row has an error in the second bit, etc. At the end of the sixth row, we have used all the single-bit error patterns, but we still have eight sequences which have not been written into the array. An arbitrary choice of one is made to head row 7 and is used as an error pattern to generate the rest of that row. Then one of the remaining four patterns is chosen to head the last row, and using it as an error pattern the final row is completed.

The error patterns which make up the last two rows of the standard array in our example would not normally be considered to be correctable errors. The decoder may therefore be designed merely to detect such errors and not attempt correction. Decoding is then said to be incomplete because not all received sequences are decoded. In this respect codes in general will differ from Hamming codes which do not have any sequences falling more than one bit different from a code sequence, and for which error correction will always involve complete decoding.

Note that because the first sequence in each row is treated as an error pattern and applied to every column, the same syndrome will be obtained for every sequence in the row. When we receive a sequence we only need to know the row in which it falls and the error pattern which heads that row. The elements of a row are called a coset, and the error pattern is called the coset leader. To carry out decoding the syndrome acts as an indicator of the coset in which the received sequence falls.

For the code in question

$$\mathbf{G} = \begin{bmatrix} 1 & 0 & 1 & 0 & 1 \\ 0 & 1 & 0 & 1 & 1 \end{bmatrix}$$

$$\mathbf{H} = \begin{bmatrix} 1 & 0 & 1 & 0 & 0 \\ 0 & 1 & 0 & 1 & 0 \\ 1 & 1 & 0 & 0 & 1 \end{bmatrix}$$

We can therefore construct the syndromes for the coset leaders, treating each in turn as a received sequence. For example the syndrome of 10000 is the leftmost column of the parity check matrix, i.e. 101; the syndrome of 11000 is the sum of the two leftmost columns, i.e. 110. We find that the syndromes of the coset leaders are 000, 101, 011, 100, 010, 001, 110 and 111, respectively.

For any code, the number of syndromes cannot be less than the number of correctable error patterns. This gives us an expression for a binary code which can detect and correct t errors:

$$2^{n-k} \geq \sum_{m=0}^{t} \begin{bmatrix} n \\ m \end{bmatrix} \tag{3.6}$$

This is called the Hamming bound, and any code which meets it with equality is called a perfect code because decoding up to the limits imposed by minimum distance produces a complete decoder. The only nontrivial binary perfect codes are the Hamming codes and the Golay code which has $n = 23$, $k = 12$, $d_{min} = 7$.

3.10 CODEC DESIGN FOR LINEAR BLOCK CODES

We have now seen all the principles which can be used to design practical encoders and decoders for short block codes. By analogy to the terminology which decrees that a modulator/demodulator is a modem, an encoder/decoder is generally called a codec. Our codec design will focus on fairly short and simple block codes to keep the complexity to a minimum. It will become clear that technological limitations on complexity will limit the usefulness of the approach.

Although minimizing complexity is obviously desirable, defining it is more diffi-cult. The amount of hardware needed is one measure, the length of time to complete the calculations is another. In general there will be trade-offs between these two components and a common measure of decoder complexity (usually far more com-plex than the encoder) is the product of *hardware complexity* and *decoding delay*. The decoder designs considered here are aimed primarily at minimizing decoding delay. Later chapters will discuss codes whose structures are different, allowing different decoding methods and, reduced hardware complexity at the expense of increased delay.

The encoder requires storage for k sequences of $n - k$ parity bits, an addressing mechanism to select the appropriate sequences, a register $n - k$ bits wide and associated logic to compute the parity checks and possibly some buffering for the information bits prior to encoding. In the case of a systematic code it may be possible to dispense with the buffering by transmitting each bit at the same time as it triggers the encoding logic and then sending the parity bits immediately following the information. There is, however, a need to balance the rate of arrival of information with the transmitted bit rate, which is higher because of the extra redundant bits. This will usually entail some sort of buffering in the system.

A possible encoder arrangement is illustrated in Figure 3.1. The k parity sequences are held in the first k locations of a read-only memory, and it is assumed that the ROM can allow access to $n - k$ bits at once. The addressing is done by means of a counter which counts from 0 to $k - 1$, the increment being triggered by the arrival of a bit of information. The read enable of the ROM is considered to be positive. The register in which the final parity check sequence is computed must allow an incoming sequence to be bit-by-bit EXORed with the contents of the register and it must be possible to clear the register at the start of each new word.

The decoder will contain some elements which are virtually identical to the encoder. The formation of the syndrome will be by a slight modification of the encoder in which the final stage is to EXOR the recalculated parity bits with the received parity bits. Buffering of the received sequence will certainly be required while the decoding is carried out. The final stage will be to use the syndrome to access a stored error pattern. At worst we shall need 2^{n-k} locations for error patterns, each of n bits, although things may be simplified if conditions such as error-free reception

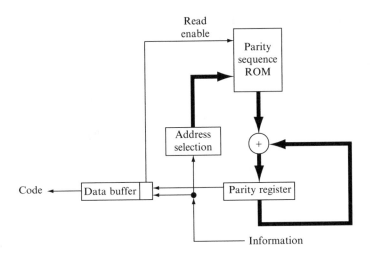

Figure 3.1 Encoder structure

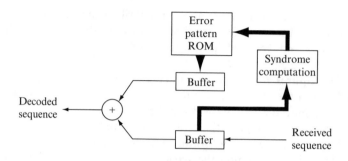

Figure 3.2   Error correction for linear block code

or uncorrectable errors are detected and handled separately. A schematic diagram for the error correction stage is shown in Figure 3.2.

These implementations are not of course the only possible ones. It would be possible for example, to construct the parity check matrix in a way that allows the error pattern to be defined relatively easily from the syndrome. It is also possible to do encoding and syndrome formation by means of buffering and a large number of hard-wired EXOR gates. Note, however, that the complexity of the decoder is going to increase as codes get longer (increasing number and size of correctable error patterns for a given value of t), and as the error-correcting capability (t) of the code increases (increasing number of correctable error patterns). Because read-only memory is relatively cheap, the more complex codes are likely to be implemented with a minimum of logic and a maximum of memory.

3.11   MODIFICATIONS TO BLOCK CODES

To confine our attention to Hamming codes would rather limit our coding capabilities. For one thing they are only single-error-correcting. Secondly, they have only a limited range of values of length n and dimension k, and the available values may not suit the system. These problems may be overcome by looking to other types of codes, but the ones worth considering will best be left to the next chapter. There are also simple modifications that may be carried out to Hamming codes (and to other block codes). In particular, reduced values of k may be used in the codes, and it is possible to create codes with $d_{min} = 4$, i.e. single-error-correcting, double-error-detecting (SECDED) codes. SECDED codes are commonly used in computer memory protection schemes.

Expanded codes

Expanding a code means adding extra parity checks to it, i.e. increasing n while keeping k the same. In particular if we add one overall parity check to a code of odd minimum distance, then the minimum distance is increased by 1.

Considering the (7, 4) Hamming code, there is one codeword of weight zero (as always in linear codes), seven of weight 3, seven of weight 4 and one of weight 7. If we add an overall parity check to create a (8, 4) code then all codewords must become even-weight sequences. The sixteen codewords will thus become one of weight zero, fourteen of weight 4 and one of weight 8. The minimum distance of the expanded code is therefore 4. Some thought about this process will show that this increase in minimum distance will always occur when d_{min} has an odd value.

Shortened codes

Shortening a code means reducing the number of information bits, keeping the number of parity checks the same. The length n and the dimension k are thus reduced by the same amount. The way in which this is done is to set one of the information bits permanently to zero and then remove that bit from the code.

Suppose we take as our example the (7, 4) Hamming code for which

$$\mathbf{G} = \begin{bmatrix} 1 & 0 & 0 & 0 & 1 & 1 & 0 \\ 0 & 1 & 0 & 0 & 1 & 0 & 1 \\ 0 & 0 & 1 & 0 & 0 & 1 & 1 \\ 0 & 0 & 0 & 1 & 1 & 1 & 1 \end{bmatrix}$$

and

$$\mathbf{H} = \begin{bmatrix} 1 & 1 & 0 & 1 & 1 & 0 & 0 \\ 1 & 0 & 1 & 1 & 0 & 1 & 0 \\ 0 & 1 & 1 & 1 & 0 & 0 & 1 \end{bmatrix}$$

The effect of setting to zero, say, the third bit of information would be to remove the third row from consideration in the generator matrix

$$\mathbf{G} = \begin{bmatrix} 1 & 0 & 0 & 0 & 1 & 1 & 0 \\ 0 & 1 & 0 & 0 & 1 & 0 & 1 \\ 0 & 0 & 0 & 1 & 1 & 1 & 1 \end{bmatrix}$$

and then to delete that bit, delete the third column

$$\mathbf{G} = \begin{bmatrix} 1 & 0 & 0 & 1 & 1 & 0 \\ 0 & 1 & 0 & 1 & 0 & 1 \\ 0 & 0 & 1 & 1 & 1 & 1 \end{bmatrix}$$

The parity checks at the end of the deleted row of the generator matrix appear as the third column of the parity check matrix, and so in the parity check matrix the third column should be deleted

$$\mathbf{H} = \begin{bmatrix} 1 & 1 & 1 & 1 & 0 & 0 \\ 1 & 0 & 1 & 0 & 1 & 0 \\ 0 & 1 & 1 & 0 & 0 & 1 \end{bmatrix}$$

We have now created a (6, 3) code; the important question is whether we have altered the minimum distance. A simple argument suffices to show that the minimum distance is not reduced; by forcing one of the information bits to zero we have reduced the number of codewords, but all the remaining codewords are still part of the original code. The minimum distance, therefore, cannot have been reduced and may have been increased by the removal of certain codewords. Neither can the deletion of one bit have had any effect on distance, because it was a zero that was deleted.

Increasing minimum distance by shortening

Assuming that we took the (7, 4) Hamming code and shortened it by deleting all the odd-weight codewords, we would then have created a code with even d_{min}; in fact, d_{min} would be 4 because we would be left with the weight 4 codewords. This could be achieved fairly easily by removing all the information bits that generate an even number of parity checks. This is equivalent to removing all the even weight columns of the parity check matrix. Thus we change

$$\mathbf{H} = \begin{bmatrix} 1 & 1 & 0 & 1 & 1 & 0 & 0 \\ 1 & 0 & 1 & 1 & 0 & 1 & 0 \\ 0 & 1 & 1 & 1 & 0 & 0 & 1 \end{bmatrix}$$

to

$$\mathbf{H} = \begin{bmatrix} 1 & 1 & 0 & 0 \\ 1 & 0 & 1 & 0 \\ 1 & 0 & 0 & 1 \end{bmatrix}$$

This is now a (4, 1) code; the information bit is repeated three times to make up the four bits of the code. Although this is a trivial example, the technique can be applied to other Hamming codes, or to other families of codes with an odd value of d_{min}, to create new codes of genuine interest.

As a second example, consider the (15, 11) Hamming code created in Section 3.8, for which the parity check matrix was

$$\mathbf{H} = \begin{bmatrix} 1 & 1 & 0 & 1 & 0 & 0 & 1 & 1 & 1 & 0 & 1 & 1 & 0 & 0 & 0 \\ 1 & 0 & 1 & 0 & 1 & 0 & 1 & 1 & 0 & 1 & 1 & 0 & 1 & 0 & 0 \\ 0 & 1 & 1 & 0 & 0 & 1 & 1 & 0 & 1 & 1 & 1 & 0 & 0 & 1 & 0 \\ 0 & 0 & 0 & 1 & 1 & 1 & 0 & 1 & 1 & 1 & 1 & 0 & 0 & 0 & 1 \end{bmatrix}$$

Removing all the even-weight columns gives

$$\mathbf{H} = \begin{bmatrix} 1 & 1 & 1 & 0 & 1 & 0 & 0 & 0 \\ 1 & 1 & 0 & 1 & 0 & 1 & 0 & 0 \\ 1 & 0 & 1 & 1 & 0 & 0 & 1 & 0 \\ 0 & 1 & 1 & 1 & 0 & 0 & 0 & 1 \end{bmatrix}$$

leaving us with a (8, 4) code with $d_{min} = 4$.

3.12   DORSCH ALGORITHM DECODING

An interesting application of matrix methods for block codes is a soft-decision decoding method invented originally by Dorsch [1], although other researchers have independently formulated very similar approaches [2–5].

The algorithm uses soft-decision information to rank the reliability of the received symbols and attempts to rework either the generator or the parity check matrix in such a way that the low-reliability symbols are treated as parity checks whose correct values are defined by the high-reliability symbols. The high-reliability symbols are called the *information set* and the low-reliability symbols are called the *parity set*.

If we erase the low-reliability parity set and re-encode from the information set, it is highly likely that most of the erroneous symbols will be corrected because they will be part of the parity set. Indeed it is possible that the codeword generated will be the maximum likelihood solution. However there is still a reasonable chance that the information set may contain some errors. If we make changes to the information set and re-encode after each change, we will generate further codewords that may also represent the maximum likelihood decoding. After a certain number of these re-encoding operations, we can compare the generated code words with the received sequence and choose the closest one, using a soft-decision measure of distance for the choice.

One complication is that in the general case not all possible choices of symbols may be used as an information set. This can be checked and the information set adjusted during the recomputation of the generator or parity check matrix.

Example of Dorsch algorithm decoding

Consider the (7, 4) Hamming code whose parity check matrix is as follows:

$$\mathbf{H} = \begin{bmatrix} 1 & 1 & 1 & 0 & 1 & 0 & 0 \\ 0 & 1 & 1 & 1 & 0 & 1 & 0 \\ 1 & 1 & 0 & 1 & 0 & 0 & 1 \end{bmatrix}$$

Let us assume that the received sequence is 3 4 7 1 0 5 7. The least reliable bits are in positions 6 and 5 with received levels of 3 and 4, respectively (levels 3 and 4 are adjacent to the hard-decision threshold). We therefore start by swapping bits 6 and 0 to bring bit 6 into the parity set. The parity check matrix is now

$$\mathbf{H} = \begin{bmatrix} 0 & 1 & 1 & 0 & 1 & 0 & 1 \\ 0 & 1 & 1 & 1 & 0 & 1 & 0 \\ 1 & 1 & 0 & 1 & 0 & 0 & 1 \end{bmatrix}$$

For the systematic form of the generator matrix, we need the right hand column to be $[0\ 0\ 1]^T$ and we can achieve this by adding rows of the matrix. This will not affect the code because satisfying the parity checks implies that all linear combinations of the parity checks will also be satisfied. In this case we can add the bottom row (row 0) into the top row (row 2) to obtain

$$\mathbf{H} = \begin{bmatrix} 1 & 0 & 1 & 1 & 1 & 0 & 0 \\ 0 & 1 & 1 & 1 & 0 & 1 & 0 \\ 1 & 1 & 0 & 1 & 0 & 0 & 1 \end{bmatrix}$$

Now swap bits 5 and 1 to bring bit 5 into the parity set. The parity check matrix becomes

$$\mathbf{H} = \begin{bmatrix} 1 & 0 & 1 & 1 & 1 & 0 & 0 \\ 0 & 1 & 1 & 1 & 0 & 1 & 0 \\ 1 & 0 & 0 & 1 & 0 & 1 & 1 \end{bmatrix}$$

Adding the middle row (row 1) into the bottom (row 0) gives

$$\mathbf{H} = \begin{bmatrix} 1 & 0 & 1 & 1 & 1 & 0 & 0 \\ 0 & 1 & 1 & 1 & 0 & 1 & 0 \\ 1 & 1 & 1 & 0 & 0 & 0 & 1 \end{bmatrix}$$

Finally we attempt to bring the next least reliable bit, the original bit 1 (now in position 5) with received level 5, into the parity checks by swapping with bit 2. The parity check matrix becomes

$$\mathbf{H} = \begin{bmatrix} 1 & 1 & 1 & 1 & 0 & 0 & 0 \\ 0 & 0 & 1 & 1 & 1 & 1 & 0 \\ 1 & 0 & 1 & 0 & 1 & 0 & 1 \end{bmatrix}$$

Unfortunately this cannot be brought to systematic form because there are no parities on the top row. We therefore restore the parity check matrix to its previous value and instead swap bit 3 with bit 2 to obtain

$$\mathbf{H} = \begin{bmatrix} 1 & 0 & 1 & 1 & 1 & 0 & 0 \\ 0 & 1 & 1 & 0 & 1 & 1 & 0 \\ 1 & 1 & 1 & 0 & 0 & 0 & 1 \end{bmatrix}$$

To restore the systematic form we add row 2 into row 1 to obtain

$$\mathbf{H} = \begin{bmatrix} 1 & 0 & 1 & 1 & 1 & 0 & 0 \\ 1 & 1 & 0 & 1 & 0 & 1 & 0 \\ 1 & 1 & 1 & 0 & 0 & 0 & 1 \end{bmatrix}$$

The decoding part can now begin. The bit ordering for the above parity check matrix is 0 1 4 2 3 5 6, so the received sequence is reordered to 7 5 7 0 1 4 3. The hard-decision version of the information set is 1 1 1 0. Re-encode to 1 1 1 0 0 0 1 and make a soft-decision comparison with the received sequence. A suitable soft-decision distance for a received 3-bit quantized level r is $7 - r$ to a code bit value 1 and r to a code bit value 0. The soft-decision distances of each received bit from the re-encoded value are therefore 0 2 0 0 1 4 4, a total of 11.

Table 3.3 Example decoding attempts using Dorsch algorithm

Received	Information set	Error pattern	Re-encoding	SD distance
7 5 7 0 1 4 3	1 1 1 0	0 0 0 0	1 1 1 0 0 0 1	11
		1 0 0 0	0 1 1 0 1 1 0	19
		0 1 0 0	1 0 1 0 0 1 0	12
		0 0 1 0	1 1 0 0 1 0 0	20
		0 0 0 1	1 1 1 1 1 1 1	20

We can now continue by inverting the hard-decision values of some of the information bits. Let us assume that we try inverting each of the information bits in turn. The results are shown in Table 3.3.

Thus in this case the best solution of those attempted is the codeword 1 1 1 0 0 0 1. Putting this back into the original order we obtain 1 0 1 0 0 1 1 as the transmitted codeword. Note that the received sequence 3 4 7 1 0 5 7 would be hard-decision quantized to 0 1 1 0 0 1 1 which differs by two bits from the final codeword. A hard-decision decoder would therefore not have obtained this solution.

3.13 CONCLUSION

This chapter has covered the basics of block codes and many more aspects will be described in the later chapters. The most common structural feature used to assist algebraic decoding is the cyclic property and that will be explained in Chapter 4. To understand the design and decoding of the main family of cyclic codes, BCH codes, it will be necessary to study finite field arithmetic and that subject is addressed in Chapter 5. Chapter 6 then deals with BCH codes and Chapter 7 with Reed Solomon codes, which are nonbinary codes of the BCH family. Issues of what is possible with block codes and the performance they give are discussed in Chapter 8. Multistage constructions of block codes are in Chapter 9 and this includes the concept of a trellis as encountered with convolutional codes; a trellis-based sequential decoding approach is used in the view of Dorsch algorithm decoding presented in [2]. Iterative decoding for block codes is in chapter 10.

Other references will be deferred until the appropriate later chapters. However, block codes are treated in almost all of the available text books, including [6–10].

3.14 EXERCISES

1 An 8-bit byte is constructed by taking 7 information bits and adding a parity check to produce an odd number of 1s in the byte (odd parity). Is this a linear code? What are the values of n, k and minimum distance?

2 Below is given a generator matrix in systematic form. What are the values of n and k for the code? What is the parity check matrix?

$$G = \begin{bmatrix} 1 & 0 & 0 & 0 & 0 & 0 & 1 & 0 & 1 & 1 \\ 0 & 1 & 0 & 0 & 0 & 0 & 0 & 1 & 0 & 1 \\ 0 & 0 & 1 & 0 & 0 & 0 & 1 & 1 & 1 & 1 \\ 0 & 0 & 0 & 1 & 0 & 0 & 1 & 1 & 1 & 0 \\ 0 & 0 & 0 & 0 & 1 & 0 & 1 & 0 & 1 & 0 \\ 0 & 0 & 0 & 0 & 0 & 1 & 0 & 1 & 1 & 1 \end{bmatrix}$$

3 For the parity check matrix below, explain how to encode and form the syndrome of the received sequence. Obtain the generator matrix. What are the values of n and k for this code? What is the syndrome of the error patterns 110001100 and 001010010?

$$H = \begin{bmatrix} 0 & 1 & 1 & 0 & 1 & 1 & 0 & 0 & 0 \\ 1 & 0 & 1 & 1 & 0 & 0 & 1 & 0 & 0 \\ 1 & 1 & 1 & 0 & 1 & 0 & 0 & 1 & 0 \\ 0 & 0 & 0 & 1 & 1 & 0 & 0 & 0 & 1 \end{bmatrix}$$

4 A (6, 3) linear code is constructed as follows:

bit 2 is a parity check on bits 5 and 4
bit 1 is a parity check on bits 4 and 3
bit 0 is a parity check on bits 5 and 3

Find the generator and parity check matrices and the minimum distance for the code.

   Construct a standard array for the code. Determine the syndromes of the coset leaders. For each bit of any 6-bit sequence, determine a logical function of the syndrome that will indicate whether that bit is in error. Assume complete decoding by your standard array.

5 A (16, 9) linear code is constructed as follows. The information is held in bits 15, 14, 13, 11, 10, 9, 7, 6 and 5. Bit 12 is a parity check on bits 15–13, bit 8 checks bits 11–9 and bit 4 checks bits 7–5. Bit 3 is a parity check on bits 15, 11 and 7, bit 2 checks bits 14, 10 and 6, bit 1 checks bits 13, 9 and 5, bit 0 checks bits 12, 8 and 4. Obtain the parity check matrix and show that the code would be unchanged if bit 0 were calculated as a parity check on bits 3–1.

   Show that the code can be represented as a 4×4 array with information in a 3×3 array and parity checks on each row and column. Hence deduce an approach to decoding single errors. Find the minimum distance of the code.

6 You are given 15 coins which should be of equal weight, but you are told that there may be one which is different from the others. You also have a balance on which to compare the weights of different coins or sets of coins. Devise a scheme using four balance checks to determine which coin, if any, has the wrong weight.

7 What is the longest SECDED code that can be created by shortening a (31, 26) Hamming code? What is its parity check matrix?

8   Write down a parity check matrix for a (7, 4) Hamming code. Construct the parity check matrix of the expanded code by appending a zero to each row and creating an extra row representing the action of the overall parity check. Now construct the generator matrix for the original code, append an overall parity check to each row and hence obtain the parity check matrix for the expanded code. Reconcile the two forms of the parity check matrix.

9   Could the (8, 4) code created by expanding a (7, 4) Hamming code also be created by shortening a longer Hamming code?

3.15   REFERENCES

1   B.G. Dorsch, *A decoding algorithm for binary block codes and J-ary output channel*, IEEE Trans Inf Theory, Vol. IT-20 (3), pp. 391–394, 1974.

2   G. Battail, *Décodage pondéré optimal des codes linéaires en blocs – 1 Emploi simplifié du diagramme du treillis*, Annales des Télécommunications, Vol. 38, Nos. 11–12, Nov–Dec 1983.

3   G. Battail, *Décodage pondéré optimal des codes linéaires en blocs – 2 Analyse et résultants de simulation*, Annales des Télécommunications, Vol. 41, Nos. 11–12, Nov–Dec 1986.

4   M.P.C. Fossorier and S. Lin, *Soft decision decoding of linear block codes based on ordered statistics*, IEEE Trans Inf Theory, Vol. IT-41 (5), pp. 1379–1396, 1995.

5   M.P.C. Fossorier and S. Lin, *Computationally efficient soft decision decoding of linear block codes based on ordered statistics*, IEEE Trans Inf Theory, Vol. IT-42 (3), pp. 738–750, 1996.

6   S. Lin and D.J. Costello, *Error Control Coding: fundamentals and applications*, Prentice Hall, 1983.

7   G.C. Clark and J.B. Cain, *Error-Correction Coding for Digital Communications*, Plenum Press, 1981.

8   A.M. Michelson and A.H. Levesque, *Error-Control Techniques for Digital Communication*, John Wiley & Sons, 1985.

9   S.B. Wicker, *Error Control Systems for Digital Communication and Storage*, Prentice Hall, 1994.

10 M. Bossert, *Channel Coding for Telecommunications*, John Wiley & Sons, 1999.

4
Cyclic codes

4.1 INTRODUCTION

Chapter 3 showed that the properties of linearity could be used to simplify the tasks of encoding and decoding linear block codes. There are many other ways in which the structure of a code can be used to assist its implementation, and for block codes the most common structure to be encountered belongs to the subclass known as cyclic codes. Their popularity is partly because their structural properties provide protection against bursty errors in addition to simplifying the logic required for encoding and decoding, although the simplified decoding may be achieved at the expense of delay.

To obtain the best understanding of this chapter, the reader should first be familiar with the material of Sections 3.1–3.5. In addition, the concept of a syndrome defined in Section 3.8 and the ability to modify block codes as in Section 3.11 will also reappear and, although they will be explained fully in this chapter, familiarity with the appropriate material from Chapter 3 will no doubt help.

This chapter draws attention to the parallel between the polynomials which are commonly used to represent cyclic code sequences and the z-transform used in digital signal processing for the representation of sampled signals and digital filters. It would be of considerable advantage to the reader to be familiar with the idea of the z-transform and, in particular, the representation of convolution; almost any book on digital signal processing will have the necessary material.

4.2 DEFINITION OF A CYCLIC CODE

Cyclic codes are a subset of linear block codes, that is to say that we are dealing still with block codes and that all the properties of linearity and the associated techniques apply equally to cyclic codes. The cyclic property is an additional property which may be of use in many circumstances.

The structure of a cyclic code is such that if any codeword is shifted cyclically, the result is also a codeword. This does not mean that all codewords can be produced by shifting a single codeword; it does however mean that all codewords can be generated from a single sequence by the processes of shifting (from the cyclic property) and addition (from the property of linearity).

4.3   EXAMPLE OF A CYCLIC CODE

The properties of the sequences which can be used to generate cyclic codes will be stated in the next section, but for the purposes of an example we shall use a particular result, namely that it is possible to generate a cyclic code of length 7 from a generator sequence of 0001011. Bearing in mind that the all-zero sequence is always a codeword of a linear code, we may construct all the codewords as follows:

1	all-zero	0000000
2	generator sequence	0001011
3	shift generator left	0010110
4	2nd shift	0101100
5	3rd shift	1011000
6	4th shift	0110001
7	5th shift	1100010
8	6th shift	1000101
9	sequences 2 + 3	0011101
10	shift sequence 9	0111010
11	2nd shift	1110100
12	3rd shift	1101001
13	4th shift	1010011
14	5th shift	0100111
15	6th shift	1001110
16	sequences 2 + 11	1111111

What we have done here is to start from the generator sequence and shift it cyclically left until all seven positions have been registered. We then find two of those sequences which add together to give a new sequence, and then shift cyclically left again until a further seven sequences have been generated. It is then found that there are two sequences which add to form 1111111, which remains the same if shifts are applied. Further shifts and additions will not create any more code sequences.

As there are 16 codewords in the above code we have four bits of information, and thus a (7, 4) code. The minimum distance can be seen to be 3 because the minimum weight nonzero codeword has weight 3. The code has the same properties as the example code from Chapter 2 and is indeed another example of a Hamming code, this time in cyclic form (there are both cyclic and noncyclic versions of Hamming codes, depending on the ordering of the columns in the parity check matrix).

4.4   POLYNOMIAL REPRESENTATION

The special methods of encoding and decoding that apply to cyclic codes are best understood through the use of an algebra in which a polynomial is used to represent sequences. In the polynomial representation, a multiplication by X represents a shift to the left, i.e. to one position earlier in the sequence. For those familiar with

z-transforms of digital signals, there is a direct parallel with the way in which the z operator represents a unit advance in time.

The terms in a polynomial represent the positions of the ones in the sequence, the rightmost position being the X^0 position, the next left the X^1 position, the next the X^2 position, etc. The generator sequence for the above code is therefore

$$g(X) = X^3 + X + 1$$

We could have taken any of the shifted positions of this basic sequence as the generator for our code, but conventionally we always take the case where the generator is shifted as far to the low powers of X as possible.

4.5 ENCODING BY CONVOLUTION

If we take the generator sequence and its first $k - 1$ left shifts, we find that we have k linearly independent sequences, that is to say that none of them can be produced by addition of two or more of the others. For our example code, we could therefore use the properties of linearity to produce any codeword by additions of sequences selected from 1011000, 0101100, 0010110 and 0001011. This would mean the code having a generator matrix (see Section 3.6) of

$$G = \begin{bmatrix} 1 & 0 & 1 & 1 & 0 & 0 & 0 \\ 0 & 1 & 0 & 1 & 1 & 0 & 0 \\ 0 & 0 & 1 & 0 & 1 & 1 & 0 \\ 0 & 0 & 0 & 1 & 0 & 1 & 1 \end{bmatrix}$$

The sequences used to generate the code, when put into polynomial form, are all multiples of the generator polynomial. Any codeword can therefore be considered to be the product of the generator and some polynomial, this polynomial representing the information content of the codeword:

$$c(X) = g(X)i(X)$$

The information is a k-bit quantity which means that $i(X)$ has as its highest possible power of X a term in X^{k-1}. The polynomial is thus said to be of *degree* $k - 1$. As $c(X)$ is of degree $n - 1$, the degree of $g(X)$ must be $n - k$.

Another equivalent view of the code generation may be obtained from the analogy between the polynomial representation and the z-transform. It is found that multiplying two z-transforms is equivalent to *convolution* of the equivalent sequences. We can therefore view the encoding as a convolution of the information with the generator sequence. Discrete convolution is written $c(j) = a(j) \otimes b(j)$ and defined by

$$c(j) = \sum_{i=0}^{j} a(i)b(j - i) \tag{4.1}$$

where j represents the position in the sequence. Both $a(j)$ and $b(j)$ are of the same length, being padded with leading zeros to the total length of the convolution. We shall see from the following example what length that should be.

Example

Consider the case where $b(j)$ is the generator sequence 1011 (with the zero order term on the right) and $a(j)$ is the information sequence 1010. The convolution is carried out as follows:

$j = 0$ $a(0)b(0) = 0$
$j = 1$ $a(0)b(1) + a(1)b(0) = 0 \times 1 + 1 \times 1 = 1$
$j = 2$ $a(0)b(2) + a(1)b(1) + a(2)b(0) = 0 \times 0 + 1 \times 1 + 0 \times 1 = 1$
$j = 3$ $a(0)b(3) + a(1)b(2) + a(2)b(1) + a(3)b(0) = 0 \times 1 + 1 \times 0 + 0 \times 1 + 1 \times 1 = 1$
$j = 4$ $a(1)b(3) + a(2)b(2) + a(3)b(1) = 1 \times 1 + 0 \times 0 + 1 \times 1 = 0$
$j = 5$ $a(2)b(3) + a(3)b(2) = 0 \times 1 + 1 \times 0 = 0$
$j = 6$ $a(3)b(3) = 1 \times 1 = 1$

For $j = 7$ or more it is seen that the sequences cannot overlap in the convolution and so the result must be 0. The length of the convolution is therefore 7 and the codeword is 1001110, which is the same as adding the first and third rows of the generator matrix above.

It is found that convolution is commutative, i.e. $a(j) \otimes b(j) = b(j) \otimes a(j)$, and that if one sequence is of length l and the other of length m, then the length of the convolution is $l + m - 1$. Thus an information sequence of length k can be convolved with a generator sequence of length $n - k + 1$ to give a code sequence of length n.

4.6   ESTABLISHING THE CYCLIC PROPERTY

If we take the generator and shift it k times, the cyclic property means that the leftmost bit now wraps around into the right position. To achieve the wrap around, every time we get a term in X^n we must add $X^n + 1$ to move the leftmost 1 into the right-hand position. We find that this can only produce a codeword if $X^n + 1$ is a multiple of the generator.

We can formulate the problem mathematically and obtain the same result as follows:

$$g(X) X^k + X^n + 1 = g^{(k)}(X)$$

where $g^{(k)}(X)$ is the polynomial obtained by cyclically shifting the generator by k places. As this is a codeword we have

$$g(X) X^k + X^n + 1 = a(X) g(X)$$
$$X^n + 1 = [a(X) + X^k] g(X)$$

where $a(X)$ is some polynomial.

Thus the generator polynomial of a (n, k) cyclic code must be a factor of $X^n + 1$.

4.7 DEDUCING THE PROPERTIES OF A CYCLIC CODE

We have seen from the above two sections that to generate a (n, k) cyclic code, the generator must satisfy two properties:

1   Generator polynomial is a factor of $X^n + 1$.

2   Degree of generator polynomial is $n - k$.

We may also, given a generator polynomial, wish to know the properties of the code generated. It is easy to obtain the number of parity check bits from the degree of the generator, but obtaining the length is more difficult. To be able to tackle this problem, we shall need to master the art of long division in modulo-2 arithmetic.

Example

Consider the generator sequence 1011 from our previous examples. We shall carry out a long division of the sequence 10000001 (representing $X^7 + 1$) by the generator, recording only the remainders:

$$
\begin{array}{r}
1011\overline{)10000001} \\
\underline{1011} \\
1100 \\
\underline{1011} \\
1111 \\
\underline{1011} \\
1010 \\
\underline{1011} \\
0
\end{array}
$$

The modulo-2 arithmetic means that addition and subtraction are the same. It is therefore perfectly valid to subtract 1011 from sequences such as 1000; all that matters is having a 1 on the left of the subtrahend so that the degree of the remainder is reduced.

From the above example we see that the sequence 10000001 will divide exactly by 1011, i.e. $X^7 + 1$ divides by $X^3 + X + 1$. Thus the length of the code generated is 7, confirming our original assumption in Section 4.2

If we did not know in advance the length of the code generated, we could instead divide the sequence $1000\ldots0$ by the generator polynomial until the remainder is 1; we then know that if the final 0 were changed to a 1 the remainder would have been 0 and the length of the code is found.

There is one valid objection to the procedure above for finding the length of a cyclic code, namely that it finds the smallest value of n although others may be possible. For example, $X^3 + X + 1$ is also a factor of $X^{14} + 1$ and so could generate a code of length 14. Such a code, although possible, would not be very practical; the sequence $X^7 + 1$, which has weight 2, would be a codeword and thus the code would have a minimum distance of only 2. In practice, therefore, one would wish to impose another condition on any generator polynomial, namely that it is not a factor of $X^j + 1$ for any lower value of j than the desired value of n. As a result the objection raised is of no practical significance.

4.8 PRIMITIVE POLYNOMIALS

If $g(X)$ is irreducible (i.e. it has no binary factors), n is the lowest possible value such that $X^n + 1$ is a multiple of some polynomial $g(X)$ and $n = 2^{n-k} - 1$ where $n - k$ is the degree of $g(X)$, then the polynomial $g(X)$ is a generator for a Hamming code in cyclic form. The Hamming code polynomials are of wider importance in the theory of block codes and are called primitive polynomials.

Table 4.1 lists primitive polynomials of degree 8 or less. For every polynomial listed here, the polynomial representing the same bit pattern in reverse is also primitive. For

Table 4.1   Primitive polynomials

Degree	Polynomial
2	$X^2 + X + 1$
3	$X^3 + X + 1$
4	$X^4 + X + 1$
5	$X^5 + X^2 + 1$
	$X^5 + X^4 + X^3 + X^2 + 1$
	$X^5 + X^4 + X^2 + X + 1$
6	$X^6 + X + 1$
	$X^6 + X^5 + X^2 + X + 1$
	$X^6 + X^5 + X^3 + X^2 + 1$
7	$X^7 + X^3 + 1$
	$X^7 + X^3 + X^2 + X + 1$
	$X^7 + X^4 + X^3 + X^2 + 1$
	$X^7 + X^6 + X^5 + X^4 + X^2 + X + 1$
	$X^7 + X^5 + X^4 + X^3 + X^2 + X + 1$
	$X^7 + X^6 + X^4 + X^2 + 1$
	$X^7 + X + 1$
	$X^7 + X^6 + X^3 + X + 1$
	$X^7 + X^6 + X^5 + X^2 + 1$
8	$X^8 + X^4 + X^3 + X^2 + 1$
	$X^8 + X^6 + X^5 + X^3 + 1$
	$X^8 + X^7 + X^6 + X^5 + X^2 + X + 1$
	$X^8 + X^5 + X^3 + X + 1$
	$X^8 + X^6 + X^5 + X^2 + 1$
	$X^8 + X^6 + X^5 + X + 1$
	$X^8 + X^6 + X^4 + X^3 + X^2 + X + 1$
	$X^8 + X^7 + X^6 + X + 1$

example $X^4 + X + 1$ represents the pattern 10011, which means that 11001 or $X^4 + X^3 + 1$ is also primitive. The patterns with the lowest weights are the easiest to implement and thus are usually chosen for error control applications.

4.9   SYSTEMATIC ENCODING OF CYCLIC CODES

Although we have seen that encoding may be carried out by a process of convolution of the information with the generator sequence, this may not be the most convenient method because the code produced is not in systematic form, making it difficult to extract the information when decoding. We therefore wish to know whether there is a convenient method of generating cyclic codes in systematic form. It is found that there is such a method, that it is based on modulo-2 long division of sequences as encountered in Section 4.7 and that it can be conveniently implemented using shift registers with feedback.

A code in systematic form consists of the information followed by parity check bits. Applying the polynomial notation, we can shift the information into the left-most bits by multiplying by X^{n-k}, leaving a codeword of the form

$$c(X) = i(x)X^{n-k} + p(X)$$

or (remembering that addition and subtraction are the same)

$$c(X) + p(X) = i(X)X^{n-k}$$

where $i(X)$ represents the information and $p(X)$ is a sequence of parity check bits.

If we take each side modulo $g(X)$, i.e. divide by $g(X)$ and find the remainder, then, as $c(X)$ is a multiple of $g(X)$ and $p(X)$ is of lower degree than $g(X)$, we obtain

$$p(x) = i(X)X^{n-k} \bmod g(x) \qquad (4.2)$$

To encode in systematic form we therefore take $i(X)$ shifted left by $n - k$ places, divide by $g(X)$ and use the remainder as the parity checks.

Example

Consider the encoding of the sequence 0110 using 1011 as the generator sequence. We must carry out long division of the sequence 1010000 by 1011 as follows:

```
        x³ + x + 1
               1011)0110000
                    1011
                    ─────
                    1110
                    1011
                    ─────
                    1010
                    1011
                    ─────
                     001
```

The remainder is 001, which means that the codeword is 0110001. This is indeed one of the codewords generated in Section 4.3.

4.10   SYNDROME OF A CYCLIC CODE

It is fairly easy to show that if we divide a sequence by the generator and take the remainder, the result is the syndrome. To do this, consider the received sequence $r(X)$ as consisting of the sum of the code sequence $c(X)$ and an error pattern $e(X)$:

$$r(X) = c(X) + e(X)$$

Splitting both the code and the error polynomials into the information and parity positions gives

$$r(x) = i(X)X^{n-k} + p(x) + e_i(X)X^{n-k} + e_p(X)$$

where $e_i(X)X^{n-k}$ represents the error pattern in the information bits and $e_p(X)$ the errors in the parity bits.

Taking the received sequence modulo $g(X)$:

$$r(X) \bmod g(X) = \left\{ [i(X) + e_i(X)]X^{n-k} \right\} \bmod g(X) + p(X) + e_p(X)$$

In other words the remainder is the same as the parity bits recalculated from the received information plus the received parity bits. It therefore corresponds to our definition of syndrome in Section 3.8.

Example

Suppose the received sequence is 1010110. The parity sequence corresponding to information 1010 was calculated in the previous section as 011. Comparing with the received parity of 110 we see that the syndrome is 101. This can be checked by long division as below:

$$
\begin{array}{r}
1011\overline{)1010110} \\
1011 \quad\quad\quad \\
\hline
1110 \quad\quad \\
1011 \quad\quad \\
\hline
101 \quad
\end{array}
$$

The syndrome obtained is therefore 101 as expected.

4.11   IMPLEMENTATION OF ENCODING

The fundamental operation of encoding and forming a syndrome is that of division by the generator polynomial and taking the remainder. A circuit to achieve this uses

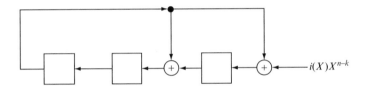

Figure 4.1   Parity calculation in encoder for cyclic (7, 4) Hamming code

an arrangement of shift register stages with feedback as shown in Figure 4.1 for the Hamming code example mentioned previously. The way that this works is to look for bits of value 1 shifted out of the left side of the registers and then to add the pattern 011 in the next three positions. This is exactly what will happen if the encoding is carried out by long division.

To see the way the circuit works, let us consider the encoding of the information 0110. The information is shifted into the registers and an extra $n - k$ shifts applied. With every incoming bit the contents of the registers are shifted left and the value shifted from the leftmost stage is modulo-2 added to the bits flowing between stages at the positions shown. At the end of this process the contents of the registers form the parity bits. The stages in the encoding are shown in Table 4.2, and the codeword is therefore 0110001 as expected. Note the fact that the register contents in the last three stages exactly correspond to the remainders produced in the example in Section 4.9.

Another way of viewing this process is to consider that the information is shifted into the registers subject to the repeated msetting of $g(X) = 0$. This means that only the remainder will be left at the end of the operation. For the example given:

$$X^3 + X + 1 = 0$$
$$X^3 = X + 1$$

Every time an X^2 term is shifted left out of the registers it becomes X^3 and is set equal to $X + 1$.

A possible criticism of the circuit of Figure 4.1 is that after the information has been entered a further $n - k$ shifts are required before the syndrome is formed. The

Table 4.2   Shift register calculation of parity checks

Input	Register contents
—	000
0	000
1	001
1	011
0	110
0	111
0	101
0	001

extra shifts can be dispensed with if the sequence is shifted into the left of the registers as shown in Figure 4.2. In this case the encoding of the information 0110 will proceed as shown in Table 4.3. The codeword is therefore 0110001 as before.

A general form of this type of circuit is shown in Figure 4.3. To determine the exact form for any given polynomial the circuit should have the same number of stages as the degree of the polynomial and the appropriate feedback connections should be made. If the shift register stages are considered as representing, from the left, the terms X^{n-k-1} down to X^0 then the positions of the connections are to the right of the stages corresponding to the terms in the polynomial. Note that it is the flow out of the leftmost X^{n-k-1} stage which corresponds to the highest power term in the polynomial.

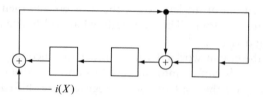

Figure 4.2   Improved method of parity calculation in encoder for cyclic (7, 4) Hamming code

Table 4.3   Improved shift register calculation of parity checks

Input	Register contents
—	000
0	000
1	011
1	101
0	001

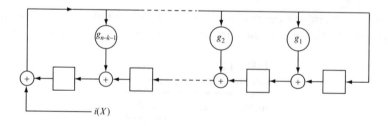

Figure 4.3   General circuit for parity calculation in encoder for cyclic codes

4.12   DECODING

The syndrome of any received sequence can be formed by shifting it into the encoder circuit of the type shown in Figure 4.1. Combinational logic or a lookup table could then be used to find the error pattern as for ordinary linear block codes. The logic can however be simplified by using the cyclic nature of the code, albeit at the expense of speed.

Suppose we form the syndrome $s(X)$ of a received sequence $r(X)$ and then apply one further shift to the registers to produce $X \cdot s(X) \bmod g(X)$. It is found that this is the same as if we had cyclically shifted the received sequence by one place and then formed the syndrome.

The syndrome depends only on the error pattern not on the transmitted codeword, and so can be found by taking the error pattern modulo $g(X)$. Hence the error pattern is the sum of the syndrome and some multiple of the generator. Thus

$$e(X) = a(X)g(X) + s(X)$$

where $e(X)$ is the error polynomial and $a(X)$ is some arbitrary polynomial. If the coefficient of X^{n-1} in $e(X)$ is e_{n-1} then the shifted error pattern is

$$e^{(1)}(X) = Xe(X) + e_{n-1}(X^n + 1)$$

$$e^{(1)}(X) = Xa(X)g(X) + e_{n-1}(X^n + 1) + Xs(X)$$

Taking the above expression modulo $g(X)$ gives the syndrome of the shifted error pattern and, as $X^n + 1$ is a multiple of $g(X)$, the remainder is just $Xs(X) \bmod g(X)$. This is the same as applying a single shift to the $s(X)$ in the syndrome registers as explained above. The result is therefore proved.

The practical significance of this result is that if we are just looking for a single error we can keep on shifting until the error reaches a chosen position (say bit $n - 1$), detect the appropriate syndrome and use the number of shifts to determine the original location of the error. Any decoder using this principle is called a Meggitt decoder.

If we take our example code and create a table of the syndromes corresponding to the possible error positions, the result is as shown in Table 4.4.

If we start from 001 and shift the registers with feedback, the result is as shown in Table 4.5. The expected relationship is apparent; if the error is for example in bit 2, then a further 4 shifts will change the syndrome into the value associated with an error in bit 6. If the error is in bit 5 then only one extra shift is needed.

Table 4.4 Syndromes for example code

Error position	Syndrome
6	101
5	111
4	110
3	011
2	100
1	010
0	001

Table 4.5   Effects of shifting syndromes

Shifts	Register contents
0	001
1	010
2	100
3	011
4	110
5	111
6	101

One does not with this method get the virtually instantaneous decoding that could be obtained using a method based purely on combinational logic or a lookup table because a number of shifts have to be applied. There is however a maximum to the number of shifts before the patterns start to repeat (at most $n - 1$ shifts are needed after formation of the syndrome).

Let us look at some examples of this principle in operation. Suppose the transmitted codeword is 1010011 but that bit 6 is received incorrectly, making the received sequence 0010011. The codeword corresponding to the received information 0010 is 0010110 (from Section 4.3) making a syndrome of 101. This can be checked either by long division or by considering the operation of the circuit of Figure 4.1. The latter is shown in Table 4.6; the former is left to the reader.

If instead bit 5 is in error, the received sequence is 1110011, the codeword with information 1110 is 1110100 and the syndrome is 111. Alternatively, looking at the operation of the circuit, we obtain the result shown in Table 4.7.

If we apply a further shift with zero at the input, the result will be 101, which is the same as the syndrome when bit 6 was in error. Thus the fact that one extra shift was required to reach the desired syndrome tells us that the error is in bit $6 - 1$, i.e. bit 5.

There is one interesting effect if we use the circuit of Figure 4.2 or Figure 4.3 for forming the syndrome. We have seen from the encoding example that using this type of circuit is equivalent to applying an extra $n - k$ shifts compared with the circuit of Figure 4.1. If bit $n - 1$ of the received sequence is in error, the effect is to form the syndrome of an error shifted left by $n - k$ positions, i.e. to position $n - k - 1$. If this

Table 4.6   Syndrome formation example

Input	Register contents
—	000
0	000
0	000
1	001
0	010
0	100
1	010
1	101

is the only error, the syndrome thus formed will be 1 followed by $n - k - 1$ zeros. The combinational logic required to detect a single-bit error in the first position thus becomes particularly simple. For our example of the received sequence 0010011 which has an error in bit 6, the syndrome formation using Figure 4.2 proceeds as shown in Table 4.8 which is the expected result. If the received sequence is 1110011 (error in bit 5) then the process is as shown in Table 4.9 and one further shift will give 100 as expected.

Table 4.7 Syndrome formation example

Input	Register contents
—	000
1	001
1	011
1	111
0	101
0	001
1	011
1	111

Table 4.8 Syndrome formation with improved encoder

Input	Register contents
—	000
0	000
0	000
1	011
0	110
0	111
1	110
1	100

Table 4.9 Syndrome formation with improved encoder

Input	Register contents
—	000
1	011
1	101
1	010
0	100
0	011
1	101
1	010

Note that although the syndrome calculated in this way is not the same as the previous definitions of syndrome, it carries the same information and therefore the term syndrome is still applied. This form of the syndrome is so common, because of its convenience, that a special term referring to it would be useful. Unfortunately there is no established term to describe it, and I shall use the symbol $\mathbf{s}^{n-k}$ or $s^{n-k}(X)$ to distinguish it, depending on whether it is treated as a vector or a polynomial.

4.13   DECODER OPERATION

The operation of a Meggitt decoder is based around the syndrome circuit of Figure 4.3. A pseudocode representation of the way in which the error detection and correction may be achieved is given below. The syntax of the pseudocode is based on the *Pascal* programming language. The code is appropriate to single-error correction, but can be extended to other cases. It includes detection of uncorrectable errors which is not required for perfect codes such as Hamming codes, but is required in all other cases.

```
begin
    shift received sequence into syndrome circuit;
    if syndrome zero then no errors
    else
        begin
        i: = n − 1;
        while syndrome <> 10 . . . 0 and i > 0 do
            begin
            i: = i − 1;
            shift syndrome register;
            end;
        if syndrome = 10 . . . 0 then error in bit i
        else uncorrectable error;
        end;
end.
```

In a practical implementation, buffering of the received sequence is required while the syndrome is formed, and at that stage it will be known whether bit $n − 1$ is in error. Bit $n − 1$ can therefore be shifted out of the buffer, corrected if necessary, and at the same time a further shift is applied to the syndrome circuit to decide whether the next bit $(n − 2)$ requires correction. In this way the data can then be shifted out of the buffer at the same time as further shifts are applied to syndrome circuit and no buffering of the error pattern is required.

4.14   MULTIPLE-ERROR CORRECTION

There are many cyclic codes which are capable of correcting multiple errors. One example is the Golay code which is a perfect (23, 12) code with $d_{min} = 7$. The generator polynomial is either

$$g(X) = X^{11} + X^{10} + X^6 + X^5 + X^4 + X^2 + 1$$

or

$$g(X) = X^{11} + X^9 + X^7 + X^6 + X^5 + X + 1$$

There are several methods of decoding such a code, but all are based on the Meggitt decoder and a straightforward implementation would be to use the circuit of Figure 4.3 and an extension of the decoding logic as follows:

1   Form the syndrome of the received sequence.

2   Look for syndromes corresponding to any correctable error patterns which include an error in bit $n - 1$.

3   If we detect such a pattern after i additional shifts, this tells us that bit $n - (i + 1)$ was in error, and so that bit can be corrected.

4   The contribution to the syndrome of any bit which has just been corrected will be a 1 in the leftmost place. The first bit in the syndrome registers is therefore inverted when any bit is corrected so that the remaining value in the syndrome registers represents the errors in the uncorrected bits.

5   Continue until the remaining value in the syndrome registers is zero or all n bits of the received sequence have been assessed.

Strictly speaking, step 4 is not necessary, but it helps in deciding when all errors have been found and in detecting conditions where uncorrectable errors have occurred.

4.15   EXAMPLE OF MULTIPLE-ERROR CORRECTION

The polynomial

$$g(X) = X^8 + X^4 + X^2 + X + 1$$

generates a (15, 7) double-error correcting code. If an encoder of the type shown in Figure 4.3 is used to form a syndrome, an error in position 14 will have a syndrome 10000000 and the syndromes of all the other single-bit errors can be found by shifting through the syndrome registers. The complete sequence is shown in Table 4.10.

The syndromes to look for are those resulting from an error in bit 14, either on its own or in combination with one other bit. This gives rise to the list shown in Table 4.11.

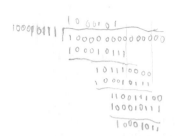

ERROR CONTROL CODING

Table 4.10 Syndromes of single errors in double-error correcting code

Error position	Syndrome
0	00010111
1	00101110
2	01011100
3	10111000
4	01100111
5	11001110
6	10001011
7	00000001
8	00000010
9	00000100
10	00001000
11	00010000
12	00100000
13	01000000
14	10000000

Table 4.11 Syndromes for double-error correction

Error positions	Syndrome
14, 0	10010111
14, 1	10101110
14, 2	11011100
14, 3	00111000
14, 4	11100111
14, 5	01001110
14, 6	00001011
14, 7	10000001
14, 8	10000010
14, 9	10000100
14, 10	10001000
14, 11	10010000
14, 12	10100000
14, 13	11000000
14	10000000

Now suppose that the errors are in positions 12 and 5. By adding the syndromes of those single errors we get a syndrome value 11101110 as being computed by the encoder. This does not appear on our list of syndromes that the encoder will try to detect, so shift once to give 11001011 and once again to give 10000001. This is on the list, the two shifts needed to reach this state indicating that bit 12 was in error. We therefore correct bit 12 and invert the leftmost bit of the syndrome to leave 00000001. A further seven shifts, making nine in all, will produce the pattern 10000000 indicating another correctable error in bit 5. Correcting this error and inverting the leftmost bit of the syndrome leaves zero, showing that the error correction is finished.

Suppose the errors are in bits 12 and 10. The syndrome calculated by using the encoder will be 00101000. Two shifts will bring this to 10100000. We correct bit 12 and invert the first bit of the syndrome, leaving 00100000. Two further shifts produce the syndrome 10000000 indicating an error in bit 10. If instead we did not bother to amend the syndrome after correcting the first error, those two further shifts would produce syndromes 01010111 and 10101110. This second value is also on our list of corrections, so we would correct the error in bit 10. The only drawbacks are that with a more powerful code to correct three or more errors, it would be difficult to know when the process had come to an end, and that after checking for errors in every bit we would not know whether all errors had been found or an uncorrectable pattern had occurred.

4.16 SHORTENED CYCLIC CODES

In common with all linear block codes, cyclic codes may be adapted to system parameters by shortening, which is the removal of a number of bits of information. In the case of cyclic codes, the most convenient method to shorten a code is to set the first few bits to zero and then not transmit those bits. The resulting codes are not, strictly speaking, cyclic, but they can be encoded and decoded using the same methods as cyclic codes because the leading zeros which have been omitted would have no effect on the formation of parity bits or of syndromes. Care must be taken however with the bit count when decoding because the Meggitt decoder will start off looking for errors in bits which have been omitted from the code. Clearly if it thinks it has found errors in any of those bits then an uncorrectable error pattern has occurred. Alternatively the arrangement of shift registers with feedback may be modified in such a way that the syndrome is effectively preshifted by the appropriate number of places so that searching for correctable errors can begin immediately.

Suppose we have a (n, k) cyclic code shortened to $(n - i, k - i)$. We receive a sequence $r(X)$ and wish to compute the syndrome of $X^j r(X)$, where j is the sum of i (number of bits removed) and $n - k$ (the usual amount by which the syndrome is preshifted). If $s_1(X)$ is the syndrome of $r(X)$ and $s_2(X)$ is the syndrome of X^j, then the required syndrome is $s_1(X)s_2(X) \bmod g(X)$. We therefore multiply the received sequence by $s_2(X) \bmod g(X)$ by feeding it into the appropriate points of the shift registers.

Consider, for example, the (15, 11) code generated by $X^4 + X + 1$, shortened to (12, 8). First we compute $X^7 \bmod g(X)$, which is found to be $X^3 + X + 1$. Now we arrange the feeding of the received sequence into the shift registers as shown in Figure 4.4, such that there is a feed into the X^3, X and 1 registers. If a sequence 100000000000 is fed into this arrangement, the sequence of register contents is as shown in Table 4.12. Thus the expected result is obtained, with the registers containing the syndrome 1000. If the first transmitted bit is in error, that fact will therefore be indicated immediately. Any other syndrome will indicate a need to shift until 1000 is obtained or the error is found to be uncorrectable.

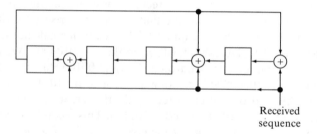

Received
sequence

Figure 4.4 Syndrome formation for shortened code

Table 4.12 Syndrome formation for shortened code

Input	Register contents
—	0000
1	1011
0	0101
0	1010
0	0111
0	1110
0	1111
0	1101
0	1001
0	0001
0	0010
0	0100
0	1000

As is the case with other linear block codes, the process of shortening cannot reduce the minimum distance and may indeed increase it. The strategies of Section 3.11 for increasing minimum distance by shortening are not, however, appropriate for implementation using the circuits designed for cyclic codes.

4.17 EXPURGATED CYCLIC CODES

Expurgation is the conversion of information bits to parity bits, i.e. keeping the length n the same, the dimension k is reduced and the number of parity symbols $n - k$ increased.

If a cyclic code has an odd value of minimum distance, multiplying the generator by $X + 1$ has the effect of expurgating the code and increasing d_{min} by 1. For example:

$$g(X) = X^3 + X + 1$$
$$g(X)(X + 1) = X^4 + X^3 + X^2 + 1$$

The degree of the new generator is increased by 1, increasing the number of parity bits; however, $X + 1$ is a factor of $X^n + 1$ for any value of n, so that the new generator is still a factor of $X^n + 1$ for the original value of n, and hence the code length is unchanged.

Any codeword of the new code consists of a codeword of the original code multiplied by $X + 1$, i.e. shifted left and added to itself. The result is bound to be of even weight because the two sequences being added are of the same weight and modulo-2 addition cannot convert even overall parity into odd. For example, taking the code sequence 1000101 from Section 4.3, shifting left and adding to itself gives

$$1\ 0\ 0\ 0\ 1\ 0\ 1 + 0\ 0\ 0\ 1\ 0\ 1\ 1 = 1\ 0\ 0\ 1\ 1\ 1\ 0.$$

Each of the sequences being added was of weight 3, but addition has caused cancellation of two of the ones leaving a codeword of weight 4.

Assuming that the original code had an odd value of minimum distance, and therefore contained odd-weight codewords, the codewords of the expurgated code are just the even-weight codewords of the original code. The term *expurgation* arises from the removal of all the odd-weight codewords. The result is to increase the minimum distance to some even value.

In the example case where the generator $X^3 + X + 1$ was expurgated to $X^4 + X^3 + X^2 + 1$, the new generator is of weight 4, so it is obvious that the new d_{min} cannot be greater than 4. Because the minimum distance must have increased to an even value from its original value of 3, it must now be exactly 4. In other cases of expurgated Hamming codes, the generator may be of higher weight, but it can still be shown that the code contains codewords of weight 4, so that after expurgation $d_{min} = 4$.

Proof

Let a code be generated by a polynomial $g(X)$ which is primitive of degree c. We choose three distinct integers p, q and r all less than $2^c - 1$ such that $X^p + X^q + X^r$ is not a codeword. If we divide this sequence by $g(X)$ we obtain a remainder $s(X)$ which, because $g(X)$ generates a perfect single-error correcting code, must be able to be interpreted as the syndrome of a single-bit error X^s. Thus there is a sequence $X^p + X^q + X^r + X^s$ which is a codeword. Moreover the integer s cannot be equal to p, q or r because that would imply the existence of a codeword of weight 2 and a minimum distance of 2. Therefore any cyclic code generated by a primitive polynomial has codewords of weight 4.

The expurgated code could be decoded in the usual way based on the new generator, but because the codewords are also codewords of the original Hamming code, we can form two syndromes based on division by the original Hamming code generator and a separate division by $X + 1$, as shown for our example case in Figure 4.5. If both syndromes are zero there are no errors. If both are nonzero we assume a single-bit error and attempt to correct it using the Hamming syndrome circuit in the usual way. If one syndrome is zero and the other nonzero we have an uncorrectable error. This method is advantageous in the detection of uncorrectable errors.

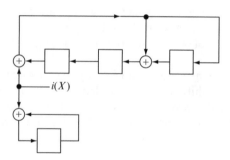

Figure 4.5   Syndrome formation for expurgated code

Example

The sequence 0111010 is a codeword of the (7, 3) expurgated code generated by $g(X) = X^4 + X^3 + X^2 + 1$. The following events give rise to the syndromes shown if the circuit of Figure 4.5 is used:

- Received sequence 0110010 (single error), syndromes are 101 and 1 (correctable error).

- Received sequence 0110011 (double error), syndromes are 110 and 0 (uncorrectable error).

- Received sequence 1011000 (triple error), syndromes are 000 and 1 (uncorrectable error).

In the first case, shifting the first syndrome gives 001, 010, 100, showing that the error is in bit 3.

4.18   BCH CODES

Many of the most important block codes for random-error correction fall into the family of BCH codes, named after their discoverers Bose, Chaudhuri and Hocquenghem. BCH codes include Hamming codes as a special case. There are binary and multilevel BCH codes, although only binary codes will be considered at the moment. For a full understanding of BCH codes, including the construction of the generator polynomial, it is necessary to have an understanding of the construction of finite fields, which will be treated in Chapter 5. Nevertheless the generator polynomials are to be found in most text books and it is easy to look them up.

The construction of a t-error correcting binary BCH code starts with an appropriate choice of length:

$$n = 2^m - 1 \quad (m \text{ is integer} \geq 3)$$

The values of k and $d_{\min}$ cannot be known for sure until the code is constructed, but one can say that

$$n - k \leq mt \quad \text{(equality holds for small } t)$$

and

$$d_{\min} \geq 2t + 1$$

where t is the design value of the number of errors to be detected and corrected. The actual code may exceed this expected value of minimum distance.

4.19 CYCLIC CODES FOR BURST-ERROR CORRECTION

There are a number of block codes which will correct single bursts within a block. Cyclic codes are generally used because of the particular properties of burst-error detection which they all possess. Not all cyclic codes, however, will possess good burst-error correction properties.

As we saw previously, a (n, k) cyclic code can be formulated in a systematic way with the $n - k$ parity check symbols in the low-order positions. The cyclic property means, however, that any consecutive $n - k$ symbols can be shifted into the parity positions and we will still have a codeword. Any error which affects only the parity symbols cannot produce a codeword result because the parity symbols are firmly fixed by the information. It therefore follows that any error that spans $n - k$ symbols or less of a cyclic codeword cannot produce a codeword result, and is therefore detectable. The cyclic nature of the code means that errors affecting the first few and last few symbols can be considered as a single end-around burst. Figure 4.6 shows an error pattern which by normal considerations would be a burst of length 14, but viewed as an end-around burst its length is only 6.

If the code is suitable for burst correction, then the maximum length of a correctable single burst within one codeword is $(n - k)/2$. This result, known as the *Reiger Bound*, may be seen by analogy with the random error case in which the error correction capability is half that of error detection, or by considering the principle of decoding when the maximum likelihood error pattern is considered to be the shortest possible burst.

The usual decoding method for burst-error correction is called error trapping, and is very similar to Meggitt decoding. Remembering that the syndrome of an error in the parity symbols is equal to the error pattern itself, we see that if the syndrome is shifted around a Meggitt decoder we will eventually reach a point where it shows directly in the syndrome. If the error pattern occupies at most $(n - k)/2$ consecutive

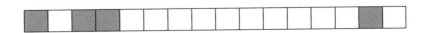

Figure 4.6   End-around burst of length 6

symbols, then we can detect this condition by the existence of $(n - k)/2$ consecutive zeros in the syndrome. The number of shifts to achieve this condition will show where the burst is located.

Example

A (15, 9) cyclic code generated by

$$g(X) = X^6 + X^5 + X^4 + X^3 + 1$$

can correct bursts of length up to 3. An error pattern

$$e(X) = X^9 + X^8 + X^7$$

has the syndrome (as formed by a circuit of the type shown in Figure 4.3)

$$s^6(X) = X^5 + X + 1$$

This pattern of 100011 when shifted gives 111111, 000111, 001110, 011100 and 111000. We now have three zeros in the low-order bits of the syndrome and the error pattern trapped in the high-order bits. The fact that it took five extra shifts to achieve this shows that the error pattern starts from bit 9, i.e. it affects bits 9, 8 and 7. We have thus correctly identified the pattern and location of the error.

Fire codes

Fire codes are cyclic codes that correct single-burst errors, with a syndrome that can be split into two components for faster decoding. The form of the generator polynomial for a Fire code correcting single bursts of length up to l is

$$g(X) = (X^{2l-1} + 1)h(X)$$

where $h(X)$ is an irreducible binary polynomial of degree $m \geq l$ which is not a factor of $X^{2l-1} + 1$, i.e. the period p of $h(X)$ is not a factor of $2l - 1$. The length of the code is the lowest common multiple of p and $2l - 1$.

As an example, let $h(X) = X^4 + X + 1$. This is a primitive polynomial of degree 4 which is therefore not a factor of $X^7 + 1$. The polynomial

$$g(X) = (X^7 + 1)(X^4 + X + 1)$$

therefore generates a (105, 94) Fire code which can correct single bursts of length 4.

We could construct a decoder in which the received sequence passes through the shift registers before feedback is applied, i.e. of the type shown in Figure 4.1 but based on $g(X)$ for the Fire code. In that case if a burst error occurred which affected bits $n + l - i - 1$ to $n - i$ (both values taken modulo n), then after a further i shifts in the registers the errors would appear in bits $l - 1$ to 0 of the syndrome. At most $n - 1$ shifts would be required to achieve this condition in which the error would be trapped and could be located.

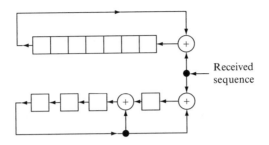

Figure 4.7   Split syndrome formation for (105, 94) Fire code

There is an alternative, faster, decoder structure in which the generator polynomial is broken down into its factors $X^{2l-1} + 1$ and $h(X)$. Such an arrangement is shown for our example code in Figure 4.7. Any error pattern which is not a codeword will leave a nonzero remainder in one or both of the syndrome registers. Clearly any error pattern of length $2l - 1$ or less cannot leave a zero remainder when divided by $X^{2l-1} + 1$ and no error pattern of length 4 or less could leave a zero remainder when divided by $h(X)$. Thus the correctable error patterns will leave a nonzero remainder in both registers.

If, with a correctable error pattern, we cycle the registers which divide by $X^{2l-1} + 1$, after λ_1 shifts the error pattern will appear in the l low order bits of the register leaving the $l - 1$ high-order bits zero. Because the period of these registers is $2l - 1$, the same pattern would reappear every $2l - 1$ further shifts and one of these occasions would correspond with the number of shifts (i) required to trap the error in a standard error trapping decoder as discussed above. Hence

$$i = A_1(2l - 1) + \lambda_1 \tag{4.3}$$

where A_1 is an unknown integer.

The next stage is to cycle the registers which divide by $h(X)$. After λ_2 shifts the error pattern appears in the registers and can be recognized because it is the same as that in the other registers. The period of $h(X)$ is p and by the same logic as before

$$i = A_2 p + \lambda_2 \tag{4.4}$$

where A_2 is an unknown integer.

Eliminating i from Equations (4.3) and (4.4) gives

$$A_1(2l - 1) - A_2 p = \lambda_2 - \lambda_1 \tag{4.5}$$

and they can also be combined to give

$$i(\lambda_2 - \lambda_1) = A_1(2l - 1)\lambda_2 - A_2 p \lambda_1 \tag{4.6}$$

We can certainly find a pair of integers A_1 and A_2 which satisfy Equation (4.5) and then substitute into (4.6). A better way, however, is to find the pair of integers I_1 and I_2 which satisfy

$$I_1(2l - 1) - I_2p = 1 \tag{4.7}$$

and then take $(\lambda_2 - \lambda_1)I_1$ and $(\lambda_2 - \lambda_1)I_2$ as the solutions of Equation (4.5) to give the following expression for Equation (4.6):

$$i = I_1(2l - 1)\lambda_2 - I_2p\lambda_1 \tag{4.8}$$

The value for i is taken modulo n. The advantage of this approach is that I_1 and I_2 can be calculated in advance and their values stored to use in Equation (4.8) once λ_1 and λ_2 are known. For example, in our case where $2l - 1$ is 7 and p is 15 the values of I_1 and I_2 are 13 and 6, respectively.

If we wish to be able to correct bursts up to length l and simultaneously to detect bursts up to length d $(d > l)$, we can construct a Fire code generator

$$g(X) = (X^c + 1)h(X)$$

where $c = l + d - 1$ and $h(X)$ is not a factor of $X^c + 1$.

4.20 CONCLUSION

As mentioned earlier, the understanding of cyclic codes is enhanced by the knowledge of finite field arithmetic, the topic of Chapter 5. BCH codes have specific properties that can be defined with the use of finite field arithmetic and they are generally decoded by algebraic means in which the syndrome is used to form a polynomial whose roots give the positions of the errors; this will be explained in Chapter 6. The use of cyclic codes as cyclic redundancy checks for error detection will be covered in Chapter 8.

In addition to the burst-error correcting codes mentioned above, there are codes known as *Burton codes* that correct phased errors, i.e. errors that fall within certain internal boundaries of the code. These are described in [1], but they can be regarded as superseded by Reed Solomon codes. The same reference also describes another variant of error trapping, Kasami decoding, for use with the Golay code; interestingly, however, there are improvements that can be made to Kasami's method [2].

4.21 EXERCISES

1   The primitive polynomial $X^4 + X + 1$ is used to generate a cyclic code. Encode the sequences 10011101001 and 01010000111.

2   Using the code of question 1, form the syndromes of the sequences 000111001110011 and 100111111110011.

3   Find the length and minimum distance of the code generated by the polynomial $X^5 + X^3 + X^2 + X + 1$.

4   Use the division method to confirm the length of the binary Golay code produced by the generator polynomial $g(X) = X^{11} + X^9 + X^7 + X^6 + X^5 + X + 1$.

5   Show that the polynomial $X^4 + X^2 + 1$ can be used to generate a code of length 12. Find the minimum distance and comment on the result.

6   Prove that an error in bit $n - 1$ of a received sequence results in a syndrome $s(X) = [g(X) + 1]/X$.

7   If the circuit of Figure 4.3 is used to form a syndrome s^{n-k} for the sequences in question 2, what will be the results? Hence determine the errors.

8   The binary polynomial $X^5 + X^2 + 1$ generates a Hamming code. What are the values of n and k? Find the syndrome value that a Meggitt decoder will attempt to find and generate an example of operation of the decoder with an error in bit 0.

9   Show that the polynomial $g(X) = X^8 + X^7 + X^6 + 1$ generates a double-error correcting code of length 15. Find the syndromes $s^8(X)$ corresponding to correctable errors including an error in bit 14. Decode the sequence 100010110010001.

10 A (15, 11) cyclic Hamming code is shortened by removal of five information bits. What are the values of length, dimension and minimum distance for the shortened code? If the generator polynomial is $g(X) = X^4 + X + 1$, encode the sequence 110001 and, using the encoder circuit to calculate syndrome, show how the decoding works if the second received bit is in error. Modify the syndrome former to premultiply by the appropriate amount and repeat the decoding process.

11 Determine the result of using the circuit of Figure 4.5 to form syndromes when there are errors

    (a) in bit 0
    (b) in bits 6 and 0

12 The binary polynomial $X^7 + X^3 + 1$ is primitive. Show that the polynomial $X^8 + X^7 + X^4 + X^3 + X + 1$ generates a (127, 119) code with $d_{min} = 4$. Design a Split syndrome decoder for the code and explain its operation.

13 Starting from the polynomial $X^3 + X + 1$, construct the generator of a Fire code of length 35 which can detect and correct all bursts of length 3 or less. Find the constants I_1 and I_2. Decode the sequence corresponding to the polynomial

$$X^{34} + X^{32} + X^{31} + X^{30} + X^{29} + X^{25} + X^{23} + X^{22} + X^{18} + X^{16} + X^{15} + X^3$$
$$+ X + 1$$

14 Amend the generator of the Fire code in Section 4.19 so that it can also detect and correct all random 2-bit errors. Find the length and dimension of the code created.

15 Which of the following codes appear to be consistent with the rules for BCH
    codes?

    $(31, 21)$ $d_{min} = 5$
    $(63, 45)$ $d_{min} = 7$
    $(63, 36)$ $d_{min} = 11$
    $(127, 103)$ $d_{min} = 7$

4.22 REFERENCES

1   S. Lin and D.J. Costello, *Error Control Coding: fundamentals and applications*, Prentice
    Hall, 1983.
2   H.P. Ho and P. Sweeney, *Cover Positions and Decoding Complexity of Golay Codes, using
    an Extended Kasami Algorithm*, IEEE Communications Letters, Vol. 2, No. 12, pp. 1–3,
    December 1998.

5

Finite field arithmetic

5.1 INTRODUCTION

So far we have confined our attention to binary arithmetic when operating on block codes. It was pointed out in Chapter 3 that binary arithmetic is a special example of the arithmetic of a finite field, and this chapter sets out to explain the general approach to finite field arithmetic. The reasons why we need to study finite field arithmetic are as follows. Firstly, we would like to understand more about the properties of cyclic codes and the design of multiple-error correcting codes. Secondly, we need to be able to implement the standard decoding methods for multiple-error correcting binary cyclic codes. Finally, we need to be able to encode and decode certain codes, principally Reed Solomon codes, that have symbols defined over a larger finite field.

There is an interesting analogy between finite field arithmetic and the more familiar topic of complex numbers that may also be of help to the reader. It will be seen that there is an interesting and useful relationship between the binary field and the larger fields derived from it, similar to that between real and complex numbers. Any irreducible binary polynomial will factorize in the appropriate larger field, just as complex numbers allow one to factorize any polynomial with real coefficients. Moreover the analogy with complex numbers will allow us to define a Fourier Transform over a finite field which is useful to envisage the encoding and decoding of certain cyclic codes.

5.2 DEFINITION OF A FINITE FIELD

What we wish to do is to have a finite set of values and some defined arithmetic operations, such that the arithmetic observes certain rules of consistency. In particular, the result of carrying out any arithmetic operation on values in the field must itself always be a member of the field. It might be thought that this would mean finding sets of real values that can be used with the usual definitions of the arithmetic operations, such that the rules are satisfied. In fact the values of the elements in the set are defined in rather an abstract way and the problem of finite field arithmetic boils down to defining the allowed operations. A finite field is also often known as a Galois field, after the French mathematician Pierre Galois. A Galois field in which the elements can take q different values is referred to as GF(q).

The formal properties of a finite field are:

(a)   There are two defined operations, namely addition and multiplication.

(b)   The result of adding or multiplying two elements from the field is always an element in the field.

(c)   One element of the field is the element zero, such that $a + 0 = a$ for any element a in the field.

(d)   One element of the field is unity, such that $a \cdot 1 = a$ for any element a in the field.

(e)   For every element a in the field, there is an additive inverse element $-a$, such that $a + (-a) = 0$. This allows the operation of subtraction to be defined as addition of the inverse.

(f)   For every non-zero element b in the field there is a multiplicative inverse element b^{-1}, such that $bb^{-1} = 1$. This allows the operation of division to be defined as multiplication by the inverse.

(g)   The associative $[a + (b + c) = (a + b) + c, a \cdot (b \cdot c) = (a \cdot b) \cdot c]$, commutative $[a + b = b + a, a \cdot b = b \cdot a]$, and distributive $[a \cdot (b + c) = a \cdot b + a \cdot c]$ laws apply.

These properties cannot be satisfied for all possible field sizes. They can, however, be satisfied if the field size is any prime number or any integer power of a prime. Our main interest will be in finite fields whose size is an integer power of 2, although to help our understanding we need first to see what happens when the field size is a prime number.

5.3   PRIME SIZE FINITE FIELD GF(p)

The rules for a finite field with a prime number (p) of elements can be satisfied by carrying out the arithmetic modulo-p. We have already seen binary arithmetic carried out modulo-2, and this satisfies all the rules for GF(2). Similarly, if we take any two elements in the range 0 to $p - 1$, and either add or multiply them, we should take the result modulo-p. The results for GF(3) are shown in Tables 5.1 and 5.2.

Table 5.1   Addition in GF(3)

+	0	1	2
0	0	1	2
1	1	2	0
2	2	0	1

Table 5.2   Multiplication in GF(3)

×	0	1	2
0	0	0	0
1	0	1	2
2	0	2	1

The additive inverse of any element is easy to identify as it is just the result of subtracting the element from p. Thus in GF(3), the additive inverse of 0 is 0, and the additive inverse of 1 is 2 and vice versa. The multiplicative inverse can in principle be found by identifying from the table pairs of elements whose product is 1. In the case of GF(3), we see that the multiplicative inverse of 1 is 1 and the multiplicative inverse of 2 is 2.

Another approach can be adopted to finding the multiplicative inverse that will be more generally useful and will lead towards the method for constructing other field sizes. In any prime size field, it can be proved that there is always at least one element whose powers constitute all the nonzero elements of the field. This element is said to be primitive. For example, in the field GF(7), the number 3 is primitive as

$$3^0 = 1$$
$$3^1 = 3$$
$$3^2 = 2$$
$$3^3 = 6$$
$$3^4 = 4$$
$$3^5 = 5$$

Higher powers of 3 just repeat the pattern as $3^6 = 1$. Note that we can carry out multiplication by adding the powers of 3, thus $6 \times 2 = 3^3 \times 3^2 = 3^5 = 5$. Hence we can find the multiplicative inverse of any element as 3^i as $3^{-i} = 3^{6-i}$. Thus in GF(7) the multiplicative inverse of 6 (3^3) is 6 (3^3), the multiplicative inverse of 4 (3^4) is 2 (3^2) and the multiplicative inverse of 5 (3^5) is 3 (3^1).

5.4   EXTENSIONS TO THE BINARY FIELD – FINITE FIELD GF(2^m)

Finite fields can also be created where the number of elements is an integer power of any prime number p. In this case it can be proved that, as was the case for GF(p), there will be a primitive element in the field and the arithmetic will be done modulo some polynomial over GF(p). In the main case of interest where $p = 2$, the polynomial to be used will be one of the primitive binary polynomials, encountered in the previous chapter as generators for Hamming codes. The use of binary polynomials means that the larger finite field inherits the property of modulo-2 addition that addition and subtraction are the same. I shall therefore ignore the formal need for minus signs in much of the mathematics in this (and the following) chapter.

Let us suppose that we wish to create a finite field GF(q) and that we are going to take a primitive element of the field and assign the symbol α to it. We cannot at present assign a numerical value to α and have to be content with the knowledge that it exists. The powers of α, α^0 to α^{q-2}, $q-1$ terms in all, form all the nonzero elements of the field. The element α^{q-1} will be equal to α^0, and higher powers of α will merely repeat the lower powers found in the finite field. The method of multiplication follows straightforwardly by modulo-($q-1$) addition of the powers of alpha. All we need now is to know how to add the powers of alpha, and this is best approached by example, through the case where $q = 2^m$ (m is an integer).

For the field GF(2^m) we know that

$$\alpha^{2^m-1} = 1$$

or

$$\alpha^{2^m-1} + 1 = 0$$

This will be satisfied if any of the factors of this polynomial are equal to zero. The factor that we choose should be irreducible, and should not be a factor of $\alpha^n + 1$ for any value of n less than $2^m - 1$; otherwise, the powers of alpha will repeat before all the nonzero field elements have been generated, i.e. alpha will not be primitive. Any polynomial that satisfies these properties is called a primitive polynomial, and it can be shown that there will always be a primitive polynomial and thus there will always be a primitive element. Moreover the degree of the primitive polynomials for GF(2^m) is always m. Tables of primitive polynomials are widely available in the literature and a selection was shown in Table 4.1.

Take as an example the field GF(2^3). The factors of $\alpha^7 + 1$ are

$$\alpha^7 + 1 = (\alpha + 1)(\alpha^3 + \alpha + 1)(\alpha^3 + \alpha^2 + 1)$$

Both the polynomials of degree 3 are primitive and so we choose, arbitrarily, the first, constructing the powers of α subject to the condition

$$\alpha^3 + \alpha + 1 = 0$$

Using the fact that each power of alpha must be α times the previous power of alpha, the nonzero elements of the field are now found to be

$$\alpha^0 = 1$$
$$\alpha^1 = \alpha$$
$$\alpha^2 = \alpha^2$$
$$\alpha^3 = \alpha + 1$$
$$\alpha^4 = \alpha \times \alpha^3 = \alpha^2 + \alpha$$
$$\alpha^5 = \alpha \times \alpha^4 = \alpha^3 + \alpha^2 = \alpha^2 + \alpha + 1$$
$$\alpha^6 = \alpha \times \alpha^5 = \alpha^3 + \alpha^2 + \alpha = \alpha^2 + 1$$

As previously stated, the nonzero powers of α can be multiplied by adding the powers of α modulo-7. To add two elements together, we must first express each element as a binary polynomial in powers of α of degree 2 or less. Addition is then by modulo-2 addition of the terms of the polynomial, for example:

$$\alpha^3 + \alpha^4 = \alpha + 1 + \alpha^2 + a = \alpha^2 + 1 = \alpha^6$$

Note that each element is its own additive inverse because of the modulo-2 addition, so that the operations of addition and subtraction are equivalent as they were in GF(2).

It may be felt that the job is not yet finished because we still do not know what α represents numerically. This is not, however, a valid objection because the numeric values are unimportant and we can assign them in any way we like. If, for example, we decide to assign the value 2 to α and 3 to α^2 then we have decided that in our arithmetic $2 \times 2 = 3$. This differs from our normal concept of arithmetic and so we might as well regard the assignment of numeric values as purely arbitrary. The process of carrying out finite field operations will therefore conceptually be as shown in Figure 5.1.

Of course there are certain mappings of values to finite field elements that may simplify the arithmetic, enabling the conversions to be integrated into the operations themselves. Later in this chapter we shall meet some mappings that are commonly adopted. However it remains true that the mapping is arbitrary and that to be able to manipulate finite fields it is sufficient to have defined the rules for multiplication and addition.

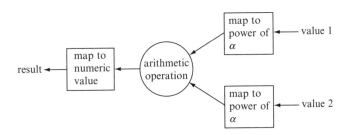

Figure 5.1 Concept of arithmetic implementation

5.5 POLYNOMIAL REPRESENTATION OF FINITE FIELD ELEMENTS

In the previous section we saw that to add two values together we need to map onto finite field elements and represent those elements as a binary polynomial in powers of α. We could incorporate this representation into the mapping to simplify the addition operation. The elements of GF(8) could then be mapped as shown in Table 5.3. The usual binary representation of any number in the range 0–7 directly represents the

Table 5.3   Polynomial mapping for GF(8)

Element	Coefficients			Value
	α^2	α^1	α^0	
0	0	0	0	0
α^0	0	0	1	1
α^1	0	1	0	2
α^2	1	0	0	4
α^3	0	1	1	3
α^4	1	1	0	6
α^5	1	1	1	7
α^6	1	0	1	5

coefficients of α^2, α^1 and α^0 and addition of two numbers can be done by modulo-2 addition of the bits.

If, for example, we wished to add 3 to 6, the binary values 011 and 110 would go bit-by-bit into exclusive-OR gates to produce the result 101, i.e. 5. This corresponds exactly to the result

$$\alpha^3 + \alpha^6 = \alpha^5$$

given in the previous section.

With this mapping, addition is easy but multiplication is more difficult. If we wish, for example, to multiply 4 by 3, we need to look up the corresponding powers of alpha, add the powers modulo-7 and then convert the result back to a numeric value as follows:

$$4 \times 3 = \alpha^2 \times \alpha^3 = \alpha^5 = 7$$

Other examples are

$$6 \times 3 = \alpha^4 \times \alpha^3 = \alpha^7 = \alpha^0 = 1$$
$$2 \times 7 = \alpha^1 \times \alpha^5 = \alpha^6 = 5$$
$$5 \times 0 = 0$$

In the last example, the rule is that any number multiplied by zero must give a zero result. In all other cases, it is seen that the process of conversion between numeric values and finite field elements has been followed, as envisaged in Figure 5.1.

We can now see that with this mapping, the process of addition is simple but the process of multiplication is difficult. There is, however, a representation of the multiplication process similar to one understood by all and very familiar before the advent of electronic calculators. If we wish to multiply two real numbers together, we can do so by taking the logarithms of the two numbers, adding the logarithms and then converting back to a real number result. In multiplying two integers under the rules of finite field arithmetic we have done a very similar thing, namely:

1   Use a table to look up the logarithm to base α of each operand.

2   Add the logarithms modulo-$q - 1$.

3   Use an antilog table to convert to the numeric value of the product.

As before we would need to trap out the case where one of the operands is zero.

We now have the situation where, with the polynomial representation, both the addition and multiplication operations can be readily understood. Addition uses bit-by-bit modulo-2 addition of the values. Multiplication uses the modulo-$q - 1$ addition of logarithms to base α. Indeed we could note that in the polynomial representation the number 2 will always be mapped onto the element α, so we could think of the log table as logarithms to base 2 – a further conceptual simplification.

Hardware implementations commonly use the polynomial representation but adopt other ideas to simplify the multiplication. Software implementations may use a mapping that makes multiplication straightforward together with an approach to reduce the complexity of addition. These matters will be discussed later in this chapter.

5.6   PROPERTIES OF POLYNOMIALS AND FINITE FIELD ELEMENTS

There are several interesting and useful relations between a field GF(p) and its extension fields GF(p^m). We will be mainly interested in the binary field and its extensions, but it may be worth bearing in mind that the principles can be extended. Many of the results will be stated and demonstrated rather than proved.

5.6.1   Roots of a polynomial

Polynomials with real coefficients do not always have real factors but always factorize if complex numbers are allowed. For example in the field of complex numbers, $X^2 + 6X + 10$ factorizes into $(X + 3 + j)(X + 3 - j)$ and the two roots are $3 + j$ and $3 - j$. These roots are said to be complex conjugates and the existence of one implies the existence of the other given that the original polynomial was real.

Similarly, an irreducible polynomial over a finite field can always be factorized in some extension field, for example the binary polynomial $X^3 + X + 1$ is found to factorize in GF(8) as $(X + \alpha)(X + \alpha^2)(X + \alpha^4)$. The values α, α^2, and α^4 are said to be the *roots* of $X^3 + X + 1$ because they represent the values of X for which the polynomial is zero.

It is easy to verify that for any binary polynomial $f(X)$, $[f(X)]^2 = f(X^2)$. In the expansion of the left-hand side, all the odd powers of X will be created an even number of times leaving a zero coefficient. Thus if β is a root of a polynomial, β^2 will also be a root, as will β^4, β^8, etc. Therefore the concept of conjugacy applies also to the roots of a binary polynomial when the roots are over a larger finite field.

The same applies to factorization of polynomials over larger finite fields. If $f(X)$ is an irreducible q-ary polynomial then it will have roots in some extension field $GF(q^m)$, i.e. the polynomial can be expressed as the product of several terms $(X + \alpha^i)$ where the terms α^i are elements of $GF(q^m)$. However in this case it is found that

$$f(X^q) = [f(X)]^q \tag{5.1}$$

Therefore, if we find one of the roots, β, then the conjugates are β^q, β^{q^2}, β^{q^3}, etc.

5.6.2   Minimum polynomial

If an irreducible polynomial $f(X)$ has β as a root, it is called the minimum polynomial of β (or of any of its other conjugate roots). If β is a primitive element then $f(X)$ is a primitive polynomial. We have already seen that the generation of a finite field is done in terms of a primitive element which is treated as a root of a primitive polynomial.

As an example, consider the finite field $GF(8)$ generated by the primitive polynomial $X^3 + X + 1$. Substituting $X = \alpha$, $X = \alpha^2$ or $X = \alpha^4$ into the polynomial gives a zero result, and the polynomial is therefore the minimum polynomial of α, α^2 and α^4. Similarly substituting α^3, α^6 or α^{12} ($\equiv \alpha^5$) into $X^3 + X^2 + 1$ verifies that they are roots. The minimum polynomial of α^0 is $X + 1$.

5.6.3   Order of an element

If m is the smallest integer value for which $\beta^m = 1$, the element β is said to be of order m and it must be a root of $X^m + 1$. If it is also a root of some irreducible polynomial $f(X)$, then $f(X)$ is a factor of $X^m + 1$. In our $GF(8)$ example, the lowest value of m for which $(\alpha^3)^m = 1$ is 7. The element α^3 is therefore of order 7, it is a factor of $X^7 + 1$ and its minimum polynomial $X^3 + X^2 + 1$ is therefore a factor of $X^7 + 1$.

5.6.4   Finite field elements as roots of a polynomial

The roots of $X^{2^c-1} + 1$ are the non-zero elements of $GF(2^c)$. Consider as an example the finite field $GF(8)$. We have already seen that the factors of $X^7 + 1$ are $X^3 + X + 1$, $X^3 + X^2 + 1$ and $X + 1$. We have also seen that α is a root of $X^3 + X + 1$; hence, α^2 and α^4 are also roots, and α^3 is a root of $X^3 + X^2 + 1$; hence, α^6 and α^5 are also roots. The root of $X + 1$ is 1. Each of the nonzero elements is therefore a root of one of the factors of $X^7 + 1$ and therefore of $X^7 + 1$ itself.

5.6.5   Roots of an irreducible polynomial

If we consider an irreducible binary polynomial of degree c, it will have c roots β, β^2, β^4, ..., $\beta^{2^{c-1}}$. Now $\beta^{2^c} = \beta$; hence,

$$\beta^{2^{c-1}} = 1$$

Thus β is a root of $X^{2^{c-1}} + 1$. As the roots of $X^{2^{c-1}} + 1$ are the nonzero elements of GF(2^c), it can be seen that an irreducible binary polynomial of degree c always has roots in GF(2^c). Conversely, the factors of $X^{2^{c-1}} + 1$ include all the irreducible polynomials of degree c. Thus $X^3 + X^2 + 1$ and $X^3 + X + 1$ are the only irreducible polynomials of degree 3.

Note that $X^m + 1$ divides into $X^n + 1$ if and only if m divides into n. This, in conjunction with the previous results, means that all irreducible polynomials of degree c are primitive if 2^{c-1} is prime. Thus because 7 is prime, all irreducible polynomials of degree 3 are primitive. On the other hand the irreducible polynomials of degree 4 are not all primitive because 15 is not a prime number.

5.6.6 Factorization of a polynomial

If we wish to factorize a binary polynomial $f(X)$, we need to establish the finite field in which the factors may be found. To do this, we first find the binary factors of the polynomial (if any) and the order of the binary polynomial representing those factors. Now find the LCM, c', of the orders; the factors of the polynomial will be found in GF ($2^{c'}$).

Proof

$$2^{ab} - 1 = (2^a)^b - 1$$

$$2^{ab} - 1 = (2^a - 1)[(2^a)^{b-1} + (2^a)^{b-2} + (2^a)^{b-3} + \cdots + 1]$$

Thus $2^a - 1$ is a factor of $2^{ab} - 1$, and so $2^{c'} - 1$ is a multiple of $2^c - 1$ if c' is a multiple of c. By choosing c' to be a multiple of the order c of some binary factor, the roots of that binary factor in GF(2^c) can also be found in GF($2^{c'}$). If c' is a multiple of the orders of all the binary factors then all the roots can be represented in GF($2^{c'}$).

As an example, the polynomial $X^5 + X^4 + 1$ factorizes into $(X^3 + X + 1)$ $(X^2 + X + 1)$. It therefore has factors in GF(2^6).

5.7 FOURIER TRANSFORM OVER A FINITE FIELD

The analogy between elements of a finite field and complex numbers can be carried to a stage further by showing that in certain circumstances we can define a discrete Fourier transform (DFT) over a finite field. Moreover it is found that the transform so defined has considerable practical value, providing us with efficient encoding and decoding techniques and an alternative view of cyclic codes which is of particular interest to those working in the field of digital signal processing. We shall start from the familiar form of the DFT, and then develop the finite field version.

The definition of the DFT of an n-point sequence in the field of complex numbers is usually expressed in terms of the relation between the frequency domain samples V_k and the time domain samples v_i:

$$V_k = \sum_{i=0}^{n-1} v_i e^{-j2\pi ik/n} \tag{5.2}$$

The term $e^{-j2\pi ik/n}$ can be taken as a set of powers of $e^{-j2\pi/n}$. However in complex representation, $e^{-j2\pi}$ is equal to 1; therefore, $e^{-j2\pi/n}$ is the nth root of 1.

In the finite field GF(2^m) there will be a transform of $v(X)$ into $V(z)$, equivalent to the Fourier transform, only if there is an nth root of 1 within the field, i.e. a term β such that $\beta^n = 1$. This will be satisfied if

$$\frac{2^m - 1}{n} = c$$

where c is some integer. It is not necessary for the polynomial being transformed to be defined over the same field as the transform; a polynomial with coefficients from GF(2^l) can be transformed over a field GF(2^m) provided m is a multiple of l. Thus, for example, a binary polynomial of length 7 could be transformed over GF(2^3) or over GF(2^6), or a polynomial of length 7 over GF(2^3) could be transformed over GF(2^6).

In the general case the coefficient V_k of the Fourier transform of a polynomial $v(X)$ can be defined as

$$V_k = \sum_{i=0}^{n-1} v_i(\alpha^c)\alpha^{eik} \tag{5.3}$$

where the term $v_i(\alpha^c)$ indicates that a coefficient β in GF(2^l) is replaced by β^c in GF(2^m). Note, however, that the value of e is defined by the ratio of $2^m - 1$ to the *length* of the transformed sequence.

The inverse DFT is defined as

$$v_i = \frac{1}{n} \sum_{k=0}^{n-1} V_k e^{-j2\pi k/n} \tag{5.4}$$

The component v_i of the inverse transform is

$$v_i = \frac{1}{n \bmod p} \sum_{k=0}^{n-1} V_k \alpha^{-cik} \tag{5.5}$$

where the symbols are defined over GF(p^m). Depending on the values produced by the inverse transform, it may then be possible to reduce the coefficients to values in a smaller field.

There will be two particular cases of interest to us. One will be the case where a binary vector of length $2^m - 1$ is transformed over $GF(2^m)$; the other is where a vector over $GF(2^m)$ of length $2^m - 1$ is transformed over its own field. In both cases $c = 1$ and

$$V_k = \sum_{i=0}^{n-1} v_i(\alpha)\alpha^{ik} \tag{5.6}$$

with the inverse transform being

$$v_i = \sum_{k=0}^{n-1} V_k\alpha^{-ik} \tag{5.7}$$

5.8   ALTERNATIVE VISUALIZATION OF FINITE FIELD FOURIER TRANSFORM

The easiest way to visualize the Fourier transform from the definition above is as follows. Regard the sequence to be transformed as a polynomial $f(X)$. Now to calculate the transform coefficient in position k, substitute $X = \alpha^k$.

Example

The sequence 0101100 is equivalent to $X^5 + X^3 + X^2$. The Fourier transform over $GF(8)$ is

$$V_0 = 1 + 1 + 1 = 1$$
$$V_1 = \alpha^5 + \alpha^3 + \alpha^2 = 0$$
$$V_2 = \alpha^3 + \alpha^6 + \alpha^4 = 0$$
$$V_3 = \alpha^1 + \alpha^2 + \alpha^6 = \alpha^3$$
$$V_4 = \alpha^6 + \alpha^5 + \alpha^1 = 0$$
$$V_5 = \alpha^4 + \alpha^1 + \alpha^3 = \alpha^5$$
$$V_6 = \alpha^2 + \alpha^4 + \alpha^5 = \alpha^6$$

Thus the transformed sequence is $\alpha^6 \ \alpha^5 \ 0 \ \alpha^3 \ 0 \ 0 \ 1$.

Note that when we double the position (modulo-n) we square the value of the transform. Thus $V_2 = V_1^2$, $V_4 = V_2^2$, $V_6 = V_3^2$ and $V_5 = V_6^2$. This is a consequence of Equation (5.1).

The inverse Fourier transform can be presented in a similar way. Regard the sequence as a polynomial $V(z)$ and substitute $z = \alpha^i$ to obtain the value of the inverse transform in position i.

Example

The sequence 0101100 is equivalent to $z^5 + z^3 + z^2$. The inverse Fourier transform over GF(8) is

$$V_0 = 1 + 1 + 1 = 1$$
$$V_1 = \alpha^2 + \alpha^4 + \alpha^5 = \alpha^6$$
$$V_2 = \alpha^4 + \alpha^1 + \alpha^3 = \alpha^5$$
$$V_3 = \alpha^6 + \alpha^5 + \alpha^1 = 0$$
$$V_4 = \alpha^1 + \alpha^2 + \alpha^6 = \alpha^3$$
$$V_5 = \alpha^3 + \alpha^6 + \alpha^4 = 0$$
$$V_6 = \alpha^5 + \alpha^3 + \alpha^2 = 0$$

Thus the transformed sequence is $0\ 0\ \alpha^3\ 0\ \alpha^5\ \alpha^6\ 1$. Note the relationship to the forward transform. The value of both transforms in position 0 is the same because the same substitution is being performed. However, for the other positions, the inverse transform in position i makes the same substitution as for the forward transform in position $n - k$. Therefore the only difference between the forward and inverse transforms is a reversing of the order of the samples, other than the one in position zero.

The same approach can be adopted for transforms of sequences over their own finite field. Consider the sequence $1\ \alpha^2\ \alpha^2\ 0\ \alpha^6\ 0$. This is equivalent to the polynomial $X^6 + \alpha^2 X^5 + \alpha X^4 + \alpha^2 X^3 + \alpha^6 X$. The Fourier transform is

$$V_0 = \alpha^0 + \alpha^2 + \alpha + \alpha^2 + \alpha^6 = \alpha^4$$
$$V_1 = \alpha^6 + \alpha^0 + \alpha^5 + \alpha^5 + \alpha^0 = \alpha^6$$
$$V_2 = \alpha^5 + \alpha^5 + \alpha^2 + \alpha^1 + \alpha^1 = \alpha^2$$
$$V_3 = \alpha^4 + \alpha^3 + \alpha^6 + \alpha^4 + \alpha^2 = \alpha^1$$
$$V_4 = \alpha^3 + \alpha^1 + \alpha^3 + \alpha^0 + \alpha^3 = 0$$
$$V_5 = \alpha^2 + \alpha^6 + \alpha^0 + \alpha^3 + \alpha^4 = \alpha^6$$
$$V_6 = \alpha^1 + \alpha^4 + \alpha^4 + \alpha^6 + \alpha^5 = 0$$

The forward transform is therefore $0\ \alpha^6\ 0\ \alpha^1\ \alpha^2\ \alpha^6\ \alpha^4$. The inverse transform of this sequence is $\alpha^6\ \alpha^2\ \alpha^1\ 0\ \alpha^6\ 0\ \alpha^4$. It has the same relationship to the forward transform as for the binary case.

Note that conjugacy constraints do not apply because the sequence to be transformed is over GF(8), not GF(2).

5.9　ROOTS AND SPECTRAL COMPONENTS

A consequence of the definition of the Fourier transform from Section 5.7 is that roots of a polynomial in the time domain are related to zero components in the transform over GF(2^m). A polynomial $v(X)$ has a root α^k if and only if the compon-

ent V_k of the transform is zero. Conversely, the polynomial $v(X)$ has a zero component v_i if and only if α^i is a root of the transform polynomial $V(z)$.

As an example, the binary sequence 0001011 is equivalent to $X^3 + X + 1$. As this
is the primitive polynomial used to generate the finite field GF(8), it is the min-
imum polynomial of α and therefore has roots α, α^2 and α^4. Thus we would expect
the frequency components V_1, V_2 and V_4 to be zero, and this is found to be the
case.

5.10   FAST FOURIER TRANSFORMS

Because of the repetition of values of α^{ik} in the definition of the Fourier transform, a
straightforward evaluation of the Fourier Transform components may multiply the
same sample value by the same α^{ik} term several times in carrying out the calculations.
For a long transform this may lead to considerable inefficiency.

This problem is well-known in the realm of complex DFTs and the usual solution
is the Cooley–Tukey fast Fourier transform. The FFT is a consequence of the
linearity and time shift properties of the DFT. Usually the number of samples is an
integer power of 2 and the procedure, properly called a radix-2 Cooley–Tukey FFT,
is as follows:

Split the series of n samples into an even series (samples 0, 2, 4, ..., $n - 1$) and an
odd series (samples 1, 3, 5, ..., n).

Take the DFT of each of the two series.

The DFT of the odd series will assume a time origin of sample 1, although the true
time origin for the whole series is at the time of sample 0. Therefore multiply the
values of the odd series by appropriate complex exponential values to align the
time origins of the odd and even series.

Fourier-transform the $n/2$ two-point series formed by a single point of the even
and odd transforms.

Each of the shorter DFTs can be evaluated by the same procedure, so that a long
DFT can be decomposed into a number of 2-point DFTs.

The same considerations apply to finite field Fourier transforms, except that the
lengths of the transforms are one less than an integer power of 2. Nevertheless, there
is no reason why transforms should not be decomposed into other basic lengths, or
even mixtures of lengths, provided the length of the transform factorizes.

The procedure for an FFT over a finite field is illustrated by an example of a length
15 sequence over GF(16). First we construct the finite field using $p(X) = X^4 + X + 1$,
a primitive binary polynomial of degree 4. The polynomial representation of the
elements is shown in Table 5.4.

The sequence to be transformed is $v = \alpha^5\ \alpha\ \alpha^3\ \alpha^4\ \alpha^{12}\ 1\ \alpha^7\ \alpha^9\ \alpha^{11}\ \alpha^6\ \alpha^{13}\ 0\ \alpha^3\ 0\ 1$.
The first step is to write the samples into a 3×5 array, so that three series of five
points have been formed. This is shown in Table 5.5.

Table 5.4　Polynomial representation of GF(16)

Element	Representation	Element	Representation
0	0000	α^7	1011
α^0	0001	α^8	0101
α^1	0010	α^9	1010
α^2	0100	α^{10}	0111
α^3	1000	α^{11}	1110
α^4	0011	α^{12}	1111
α^5	0110	α^{13}	1101
α^6	1100	α^{14}	1001

Table 5.5　15-point series written into columns

α^5	α^4	α^7	α^6	α^3
α	α^{12}	α^9	α^{13}	0
α^3	1	α^{11}	0	1

Now each row is Fourier transformed. Remember that from Equation (5.3)

$$V_k = \sum_{i=0}^{n-1} v_i \alpha^{cik}$$

where in this case $c = 3$. For example, for the top row

$$V_0 = \alpha^5 + \alpha^4 + \alpha^7 + \alpha^6 + \alpha^3 = \alpha^9$$
$$V_1 = \alpha^5\alpha^{12} + \alpha^4\alpha^9 + \alpha^7\alpha^6 + \alpha^6\alpha^3 + \alpha^3 = \alpha^2 + \alpha^{13} + \alpha^{13} + \alpha^9 + \alpha^3 = \alpha^5$$
$$V_2 = \alpha^5\alpha^{24} + \alpha^4\alpha^{18} + \alpha^7\alpha^{12} + \alpha^6\alpha^6 + \alpha^3 = \alpha^{14} + \alpha^7 + \alpha^4 + \alpha^{12} + \alpha^3 = \alpha^5$$
$$V_3 = \alpha^5\alpha^{36} + \alpha^4\alpha^{27} + \alpha^7\alpha^{18} + \alpha^6\alpha^9 + \alpha^3 = \alpha^{11} + \alpha^1 + \alpha^{10} + \alpha^0 + \alpha^3 = \alpha^1$$
$$V_4 = \alpha^5\alpha^{48} + \alpha^4\alpha^{36} + \alpha^7\alpha^{24} + \alpha^6\alpha^{12} + \alpha^3 = \alpha^8 + \alpha^{10} + \alpha^1 + \alpha^3 + \alpha^3 = 0$$

The result of the row transforms is given in Table 5.6.

Next the points are multiplied by the values shown in Table 5.7, corresponding to compensation for the different time offsets. The result is shown in Table 5.8.

Finally the columns are transformed using the DFT expression above with $c = 5$. The result is shown in Table 5.9.

Table 5.6　Row transforms of 15-point sequence

0	α^1	α^5	α^5	α^9
α^2	α^8	α^{14}	α^1	α^9
α^{10}	α^{13}	α^4	α^{11}	α^5

Table 5.7 Time offset factors for 15-point transform

α^8	α^6	α^4	α^2	1
α^4	α^3	α^2	α^1	1
1	1	1	1	1

Table 5.8 Row transforms compensated for time offsets

0	α^7	α^9	α^7	α^9
α^6	α^{11}	α^1	α^2	α^9
α^{10}	α^{13}	α^4	α^{11}	α^5

Table 5.9 Fourier transform

α^8	α^{11}	α^2	α^{11}	α^6
α^{14}	α^7	α^6	α^0	α^6
α^7	α^3	α^7	α^0	α^5

This completes the transform, but the values must be read out by row. The result is therefore

$$V = \alpha^8 \ \alpha^{11} \ \alpha^2 \ \alpha^{11} \ \alpha^6 \ \alpha^{14} \ \alpha^7 \ \alpha^6 \ \alpha^0 \ \alpha^6 \ \alpha^7 \ \alpha^3 \ \alpha^7 \ \alpha^0 \ \alpha^5$$

5.11 HARDWARE MULTIPLIERS USING POLYNOMIAL BASIS

In Section 5.5 we saw that multiplication in the polynomial basis was equivalent to using logarithmic tables to base α. This requires three table lookups, an inconvenience for hardware implementation. If, however, one of the factors is known in advance, efficient multiplier circuits are easy to design.

Consider the case of multiplication in GF(8). Suppose we have an element β whose bit values are β_2, β_1, β_0. Multiplying by α would be equivalent to shifting the bits once in the registers shown in Figure 5.2.

This multiplication could be expressed in matrix terms by

$$\beta \times \alpha = [\beta_2 \ \beta_1 \ \beta_0] \begin{bmatrix} 0 & 1 & 1 \\ 1 & 0 & 0 \\ 0 & 1 & 0 \end{bmatrix}$$

Note that the top row of the matrix is just low-order coefficients of the primitive polynomial. The lower rows of the matrix contain an identity matrix followed by a zero column. This format applies to all extension fields of GF(2).

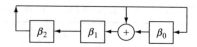

Figure 5.2   Multiplication by α in GF(8)

If we wish to multiply by α^i, we need to raise the above matrix to the power i. The results of doing this are

$$\alpha^2 \equiv \begin{bmatrix} 1 & 1 & 0 \\ 0 & 1 & 1 \\ 1 & 0 & 0 \end{bmatrix}$$

$$\alpha^3 \equiv \begin{bmatrix} 1 & 1 & 1 \\ 1 & 1 & 0 \\ 0 & 1 & 1 \end{bmatrix}$$

$$\alpha^4 \equiv \begin{bmatrix} 1 & 0 & 1 \\ 1 & 1 & 1 \\ 1 & 1 & 0 \end{bmatrix}$$

$$\alpha^5 \equiv \begin{bmatrix} 0 & 0 & 1 \\ 1 & 0 & 1 \\ 1 & 1 & 1 \end{bmatrix}$$

$$\alpha^5 \equiv \begin{bmatrix} 0 & 1 & 0 \\ 0 & 0 & 1 \\ 1 & 0 & 1 \end{bmatrix}$$

In fact, looking at any one of these we can see that the format is

$$\alpha^i \equiv \begin{bmatrix} \alpha^{i+2} \\ \alpha^{i+1} \\ \alpha^i \end{bmatrix}$$

so that each row of the matrix consists of the polynomial representation of the appropriate power of α.

Each column of the multiplication matrix denotes the bits that are added together to form a single bit of the product. For example, the α^3 multiplier could be implemented as in Figure 5.3.

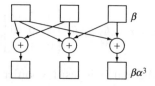

Figure 5.3   Multiplication by α^3

5.12 HARDWARE MULTIPLICATION USING DUAL BASIS

When neither of the factors in a multiplication is known in advance, the polynomial basis on its own does not provide for efficient multipliers. However if one of the factors is put into another basis, known as the *dual basis*, an arrangement known as a Berlekamp multiplier can be used.

The simplest definition of the dual basis is that for any element β in $GF(2^m)$ it consists of the least significant bits from the polynomial basis representation of $\beta, \beta\alpha, \ldots, \beta\alpha^{m-2}, \beta\alpha^{m-1}$. For the elements of GF(8), the polynomial and dual basis representations are shown in Table 5.10.

A Berlekamp multiplier over GF(8) is shown in Figure 5.4. This is a bit-serial multiplier, i.e. it produces the bits of the product (in dual basis representation) in series. The feedback connections in reverse order are based on the primitive polynomial used to generate GF(8), i.e. $X^3 + X + 1$. Shifting the dual basis representation of an element β produces the dual basis representation of the element $\beta\alpha$.

As an example, if we wish to multiply α^4 by α^2, we can put α^2 into dual basis to become 010. At the subsequent stages shifting the dual basis register gives 101 and 011 (the dual basis representations of α^3 and α^4). The results of the multiplications as shown in Table 5.11.

Table 5.10   Dual basis for GF(8)

Element	Polynomial basis	Dual basis
α^0	001	100
α^1	010	001
α^2	100	010
α^3	011	101
α^4	110	011
α^5	111	111
α^6	101	110

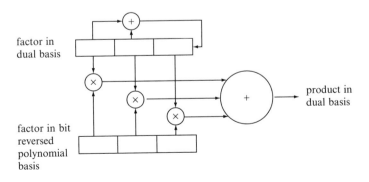

Figure 5.4   Berlekamp multiplier over GF(8)

Table 5.11 Berlekamp multiplication

Stage	Bit reversed polynomial	Dual	Product
1	011	010	1
2	011	101	1
3	011	011	0

Note that 110 is the dual basis representation of α^6.

Basis conversion

Of course the suitability of the Berlekamp multiplier for hardware implementation depends on how easy it is to convert between the polynomial basis and the dual basis. In the example above, the conversion is easy because if the polynomial representation of an element β is β_2, β_1, β_0 the dual basis is β_0, β_2, β_1. In other words it is produced by a simple reordering of the bits. Indeed it can be shown that for any finite field GF(2^m) where the primitive polynomial is a trinomial (i.e. it contains 3 terms X^m, X^c and 1), the dual basis is β_{c-1}, β_{c-2}, $\ldots$, β_0, β_{m-1}, β_{m-2}, $\ldots$, β_c.

The main field of interest for multipliers is GF(256) for which there is no primitive polynomial that is a trinomial. The usual primitive polynomial to use is a pentanomial, $X^8 + X^4 + X^3 + X^2 + 1$. The circuit for shifting the dual basis element is therefore as shown in Figure 5.5.

Note also that the mapping from the polynomial representation of element β does not need to correspond exactly to the definition of dual basis given above. In particular it could be mapped to the dual basis of an element $\beta\alpha^i$. The result of any multiplication would similarly be increased by the α^i, but the inverse mapping to polynomial basis would take care of that.

It is found that if the polynomial basis of β is mapped to the dual basis of $\beta\alpha^{-2}$, a convenient conversion exists. In this case the dual basis is

$$\beta_2 + \beta_0, \beta_1, \beta_0, \beta_7, \beta_6, \beta_5, \beta_4, \beta_3 + \beta_7$$

and the conversion of a dual basis element γ to polynomial basis is

$$\gamma_2, \gamma_1, \gamma_0 + \gamma_2, \gamma_3 + \gamma_7, \gamma_6, \gamma_5, \gamma_4, \gamma_3$$

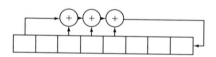

Figure 5.5 Dual basis element shifting in GF(256) multiplier

Multiplicative inverses

For any nonzero element β in GF(2^m), $\beta^{-1} = \beta^{2^m - 2}$. However $2^m - 2 = 2 + 2^2 + 2^3 + \cdots + 2^{2^{m-1}}$. Therefore

$$\beta^{-1} = \beta^2 \times \beta^4 \times \beta^8 \times \cdots \times \beta^{2^{m-1}} \tag{5.8}$$

calculation of the multiplicative inverse (for division operations) therefore involves $m - 1$ squarings and $m - 2$ multiplications. Alternatively, a lookup table of inverse values can be held in ROM.

5.13 HARDWARE MULTIPLICATION USING NORMAL BASIS

Another basis in which hardware multipliers are sometimes implemented is the *normal basis*. The normal basis of GF(2^m) is a polynomial in $\beta^{2^{m-1}}, \beta^{2^{m-2}}, \ldots, \beta^4, \beta^2, \beta$ where β is an element of order m such that the elements $\beta^{2^{m-1}}, \beta^{2^{m-2}}, \ldots, \beta^4, \beta^2, \beta$ are linearly independent. For example, in GF(8) we cannot use $\alpha^4, \alpha^2, \alpha^1$ because they are not linearly independent. However we can use $\alpha^5, \alpha^6, \alpha^3$ as our basis, the resulting representation being shown in Table 5.12.

The implementation of addition can be carried out by modulo-2 addition of the bits of the symbol, as for the polynomial basis. The particular attraction of the normal basis is that there is a certain regularity to the circuits for multiplication. For example, the result of multiplying α^3 by α^6 is $\alpha^5 + \alpha^3$; if we cyclically shift the multiplicands and the result to $\alpha^6 \times \alpha^5 = \alpha^3 + \alpha^6$, we also get a correct expression. We may therefore implement the arithmetic as shown in Figure 5.6.

The multiplicands are loaded into the registers and the accumulator is cleared. The results of multiplying the first multiplicand by the α^3 polynomial of the second are evaluated and summed into the accumulator. Now the registers and the accumulator are cyclically shifted and the process is repeated. After the third cycle, a final shift of the accumulator gives the result. The process of multiplying α^4 by α^1 is shown in Table 5.13. The final result, 100, is the normal basis representation of α^5.

Table 5.12 Normal basis for GF(8)

Element	α^5	α^6	α^3
0	0	0	0
α^0	1	1	1
α^1	1	1	0
α^2	1	0	1
α^3	0	0	1
α^4	0	1	1
α^5	1	0	0
α^6	0	1	0

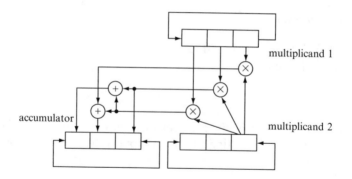

Figure 5.6   Normal basis multiplier for GF(8)

Table 5.13   Normal basis multiplication example

Start accumulator	Multiplicand 1	Multiplicand 2	Partial result	End accumulator
000	011	110	000	000
000	101	011	100	100
010	110	101	011	001
100				

If all arithmetic is implemented in the normal basis, no basis conversions are required as all bit patterns are simply regarded as a polynomial in that basis. Note also that inversion process shown in Equation (5.8) is easier to implement in the normal basis because the squaring operation is simple.

5.14   SOFTWARE IMPLEMENTATION OF FINITE FIELD ARITHMETIC

One alternative to using a polynomial representation would be to let the element value directly represent the power of α. This would solve the multiplication problem, but would create a problem with addition. We would need to look up the polynomial representation of each element, modulo-2 add the coefficients and then translate the resulting polynomial back to a power of α. This would have no overall benefit compared with polynomial representation; however, there is an alternative approach for addition using Zech logarithms.

Let

$$a^{Z(n)} = \alpha^n + 1 \tag{5.9}$$

then

$$\alpha^m + \alpha^n = \alpha^m(\alpha^{n-m} + 1) = \alpha^m \alpha^{Z(n-m)} = \alpha^{Z(n-m)+m} \tag{5.10}$$

Thus with a table of $Z(n)$ we can easily perform addition. For GF(8) the Zech logarithm table is as shown in Table 5.14.

Table 5.14 Zech logarithms
for GF(8)

n	$Z(n)$
1	3
2	6
3	1
4	5
5	4
6	2

Note that the value of $Z(0)$ is never needed as it only occurs if we try to add two terms which are the same, producing a zero result.

As examples using Zech logarithms:

$$\alpha^4 + \alpha^5 = a^{Z(1)+4} = \alpha^7 = 1$$
$$\alpha^1 + \alpha^6 = a^{Z(5)+1} = \alpha^5$$

It is not necessary to decide which is the larger power of alpha before carrying out the subtraction, provided the result is taken modulo-$(q - 1)$:

$$\alpha^5 + \alpha^4 = a^{Z(6)+5} = \alpha^7 = 1$$
$$\alpha^6 + \alpha^1 = a^{Z(2)+6} = \alpha^5$$

There is, however, a complication in direct representation of powers of α. Namely deciding whether the value 0 should represent the zero element or the element α^0. We could resolve this problem by letting the value $q - 1$ represent the zero element, leading to the following representation for GF(8).

$$0 = 111$$
$$1 = 000$$
$$\alpha = 001$$
$$\alpha^2 = 010$$
$$\alpha^3 = 011$$
$$\alpha^4 = 100$$
$$\alpha^5 = 101$$
$$\alpha^6 = 110$$

Addition is done using Zech logarithms. To multiply two nonzero numbers, take the modulo-$(q - 1)$ sum of the representations.

An alternative, which avoids the possible confusion between the value zero and the zero element, is to let the powers of α be represented by the one's complement of the direct representation above. For GF(8), the representation is now

$$0 = 000 \ (= 0)$$
$$1 = 111 \ (= 7)$$
$$\alpha = 110 \ (= 6)$$
$$\alpha^2 = 101 \ (= 5)$$
$$\alpha^3 = 100 \ (= 4)$$
$$\alpha^4 = 011 \ (= 3)$$
$$\alpha^5 = 010 \ (= 2)$$
$$\alpha^6 = 001 \ (= 1)$$

With this representation, the multiplication algorithm becomes a straightforward addition of the representations, with $q - 1$ being subtracted if the result exceeds $q - 1$. Alternatively, increment by 1 if there is a carryout from the most significant bit. Addition is also straightforward provided the Zech logarithm table is held in one's complement form. For example, the table for GF(8) would be as shown in Table 5.15.

Examples of multiplication in this field representation are

$$\alpha^4 \times \alpha^5 = 3 + 2 = 5 = \alpha^2$$
$$\alpha^2 \times \alpha^3 = 5 + 4 = 9 = 2 = \alpha^5$$

To add α^4 and α^5 we evaluate $3 + Z(2 - 3) = 3 + Z(6) = 7$, which is α^0. To add α^6 and α, evaluate $1 + Z(5) = 2$, which is α^5.

Table 5.15   One's complement form of Zech logarithm table

n	$Z(n)$
6	4
5	1
4	6
3	2
2	3
1	5

5.15   CONCLUSION

In this chapter we have covered both the basic theory of finite fields and implementation issues. In the basic theory, a Fourier transform over a finite field has been presented and this will allow a view of cyclic codes that will be helpful in understanding BCH codes and the multiple-error correcting algebraic decoding method presented in Chapter 6. The transform view is promoted in [1], where more information about fast algorithms can also be found. Most books have a treatment of the theory

of finite fields and [2] provides a good, comprehensible coverage, despite its avoidance of the transform approach.

Implementation issues must be considered in the context of the system being
designed as there is no perfect approach for all codes and all platforms. The main
emphasis of research is on hardware implementations to provide the high speeds for
modern communication systems [3–6].

5.16 EXERCISES

1   Use the primitive polynomial $X^3 + X^2 + 1$ to create a polynomial representation
    of the field elements of GF(8). Evaluate the products (111)·(100), (101)·(010),
    (011)·(110) and the divisions (100)/(101), (111)/(110) and (010)/(011).

2   Find the minimum polynomials for each of the nonzero elements of GF(8) as
    defined in question 1.

3   For the field GF(8) created in question 1, perform the sums $\alpha + \alpha^2$, $\alpha^5 + 1$,
    $\alpha^6 + \alpha^3$, $\alpha^4 + \alpha^5$.

4   Use the primitive polynomial $X^4 + X^3 + 1$ to construct the finite field GF(16) in
    polynomial form.

5   Find the forward and inverse Fourier transforms of each of the following over
    GF(8) as created in question 1:

    1001110
    1010011
    1111111
    0011010
    1101001

6   Find the forward and inverse Fourier transforms of each of the following over
    GF(8) as created in question 1:

    $\alpha^3\ \alpha^5\ \alpha\ \alpha^6\ 0\ 1\ \alpha$
    $\alpha^3\ \alpha^5\ \alpha\ 0\ 0\ 0\ 0$
    $\alpha^2\ \alpha^4\ 0\ \alpha^6\ \alpha^6\ \alpha^5\ \alpha^6$

7   Find the forward and inverse transforms of the binary sequence
    101100100000000 over GF(16) as created in question 4.

8   Find the normal and dual basis of GF(16) as created in question 4.

9   If $Z(n)$ represents the Zech logarithm, prove that

    $$Z(Z(n)) = n \quad \text{and} \quad Z(q - 1 - n) = Z(n) - n$$

5.17 REFERENCES

1   R.E. Blahut, *Theory and Practice of Error Control Codes*, Addison Wesley, 1983.

2   S. Lin and D.J. Costello, *Error Control Coding: fundamentals and applications*, Prentice Hall, 1983.

3   S.T.J. Fenn, M. Benaissa and D. Taylor, *GF(2m) multiplication and division over the dual basis*, IEEE Trans. Computers. Vol. 45, No. 3, pp. 319–327, 1996.

4   S.T.J. Fenn, M. Benaissa and D. Taylor, *Finite field inversion over the dual basis*, IEEE Trans. VLSI Systems, Vol. 4, pp. 134–137, 1996.

5   S.T.J. Fenn, M. Benaissa and D. Taylor, *Fast normal basis inversion in $GF(2^m)$*, IEE Electronic Letters, Vol. 32 No. 17, pp. 1566–1567, 1996.

6   B. Sunar and C.K. Koç, *An Efficient Optimal Normal Basis Type II Multiplier*, IEEE Transactions on Computers, Vol. 50, No. 1, pp. 83–87, January 2001.

6

BCH codes

6.1 INTRODUCTION

BCH codes are a class of cyclic codes discovered in 1959 by Hocquenghem [1] and independently in 1960 by Bose and Ray-Chaudhuri [2]. They include both binary and multilevel codes and the codes discovered in 1960 by Reed and Solomon [3] were soon recognized to be a special case of multilevel BCH codes [4]. In this chapter we shall confine our attention to binary BCH codes, leaving Reed Solomon codes to Chapter 7.

From the discussions of Chapter 4, we could, given a generator polynomial, construct an encoder for binary BCH codes. Moreover, using the discussions of finite field arithmetic in Chapter 5, it requires only a straightforward extension of the principles of Chapter 4 to construct an encoder for a Reed Solomon code, given the generator polynomial. The purpose of this chapter is to show the structure of BCH codes and the decoding methods. We will therefore be able to work out the generator polynomials for any code of interest and to implement low complexity decoders.

6.2 SPECIFYING CYCLIC CODES BY ROOTS

It is possible to specify a cyclic binary code by saying that the codewords are binary polynomials with specific roots in $GF(2^m)$. These roots, being common to every codeword, are inherited from the generator polynomial. Note that the concepts of conjugacy will apply and that the existence of a particular root will imply the existence of the conjugates. Thus the generator polynomial will be constructed from the minimum polynomials of the roots. For example if the specified root is α from GF(8), we know that the minimum polynomial is $X^3 + X + 1$, and all codewords must be multiples of this minimum polynomial. In this case the minimum polynomial acts as the generator for the code.

In general the generator polynomial will be the least common multiple of the minimum polynomials for the specified roots. The degree of the polynomial, which is equal to the number of parity check symbols for the code, is the same as the number of separate roots, so that the total number of code roots gives the number of parity check symbols.

From the discussions of Chapter 5, we recall that the values of roots of a time domain polynomial are equivalent to the positions of zeros in the frequency domain.

We shall therefore be able to say equivalently that all codewords of a cyclic code have zeros at specific locations in the frequency domain.

6.3   DEFINITION OF A BCH CODE

A t-error correcting q-ary BCH code of length $q^m - 1$ is a cyclic code whose roots include $2t$ consecutive powers of α, the primitive element of $GF(q^m)$. There will be two main cases of interest:

- *Binary BCH codes* will consist of binary sequences of length $2^m - 1$ with roots including $2t$ consecutive powers of the primitive element of $GF(2^m)$. Alternatively, the Fourier transform over $GF(2^m)$ will contain $2t$ consecutive zeros. Note that the generator polynomial will have conjugate roots (or conjugate frequency domain zeros) in addition to the specified $2t$ values.

- *Reed Solomon codes* are the special case where $m = 1$. They therefore consist of sequences of length $q - 1$ whose roots include $2t$ consecutive powers of the primitive element of $GF(q)$. Alternatively, the Fourier transform over $GF(q)$ will contain $2t$ consecutive zeros. Note that because both the roots and the symbols are specified in $GF(q)$, the generator polynomial will have only the specified roots; there will be no conjugates. Similarly the Fourier transform of the generator sequence will be zero in only the specified $2t$ consecutive positions.

The individual codewords, being multiples of the generator, may have roots in addition to the ones specified, depending on the multiplying polynomial. However the generator itself will have only the roots (or frequency domain zeros) implied by the above, and only those roots (or frequency domain zeros) will be common to every codeword.

6.4   CONSTRUCTION OF BINARY BCH CODES

To create some realistic multiple-error correcting examples, we shall need to work with codes of length 15 or more, implying a need to construct corresponding finite fields. We therefore construct $GF(16)$ so that we can define codes of length 15. The primitive polynomial to be used is $p(X) = X^4 + X + 1$. The polynomial basis representation of the elements is shown in Table 6.1.

Single-error correcting code

Suppose we choose α^1 and α^2 as the consecutive roots for this code. We know that there will be conjugate roots, the full set of conjugates being

$$\alpha^1 \ \alpha^2 \ \alpha^4 \ \alpha^8$$

Table 6.1 GF(16)

Element	Value
0	0000
α^0	0001
α^1	0010
α^2	0100
α^3	1000
α^4	0011
α^5	0110
α^6	1100
α^7	1011
α^8	0101
α^9	1010
α^{10}	0111
α^{11}	1110
α^{12}	1111
α^{13}	1101
α^{14}	1001

This set of conjugates contains the two roots we want, therefore no others are required. The generator polynomial is

$$g(X) = \left(X + \alpha^1\right)\left(X + \alpha^2\right)\left(X + \alpha^4\right)\left(X + \alpha^8\right)$$
$$g(X) = X^4 + X^3\left(\alpha^1 + \alpha^2 + \alpha^4 + \alpha^8\right) + X^2\left(\alpha^3 + \alpha^5 + \alpha^9 + \alpha^6 + \alpha^{10} + \alpha^{12}\right)$$
$$+ X\left(\alpha^7 + \alpha^{11} + \alpha^{13} + \alpha^0\right) + \alpha^0$$
$$g(X) = X^4 + X + 1$$

Alternatively we find that the generator polynomial is the LCM of the minimum polynomials of α^1 and α^2. Both these elements have $X^4 + X + 1$ as their minimum polynomial, therefore this is the generator of the code. The code has 4 parity checks (from the degree of the generator) and so is a (15, 11) code.

Note that the generator polynomial is primitive and that therefore the code created is a Hamming code. We can therefore see that cyclic Hamming codes are just single-error correcting binary BCH codes.

Double-error correcting code

Following on from the above example, we choose α^1 α^2 α^3 and α^4 as our desired roots. Starting from α and including the conjugates we find as for the Hamming code that we have roots

$$\alpha^1 \; \alpha^2 \; \alpha^4 \; \alpha^8$$

However we still need another root, namely α^3. Including the conjugates we create roots

$$\alpha^3 \; \alpha^6 \; \alpha^{12} \; \alpha^9$$

The generator polynomial is

$$g(X) = (X + \alpha^1)(X + \alpha^2)(X + \alpha^4)(X + \alpha^8)(X + \alpha^3)(X + \alpha^6)(X + \alpha^{12})(X + \alpha^9)$$
$$g(X) = (X^4 + X + 1)(X^4 + X^3 + X^2 + X + 1)$$
$$g(X) = X^8 + X^7 + X^6 + X^4 + 1$$

Alternatively we can see that $X^4 + X + 1$ is the minimum polynomial of α^1 α^2 and α^4, the minimum polynomial of α^3 is found to be $X^4 + X^3 + X^2 + X + 1$ leading to the result above. The degree of the generator is 8, so the code is (15, 7).

Triple-error correcting code

To make a triple-error correcting code we want to have roots α^1 α^2 α^3 α^4 α^5 and α^6. Note that the double-error correcting example had all these except α^5, so we take another set of conjugate roots, namely

$$\alpha^5 \; \alpha^{10}$$

The minimum polynomial of α^5 is found to be $X^2 + X + 1$, leading to

$$g(X) = (X^8 + X^7 + X^6 + X^4 + 1)(X^2 + X + 1)$$
$$g(X) = X^{10} + X^8 + X^5 + X^4 + X^2 + X + 1$$

With a degree 10 generator, and therefore 10 parity checks, this is a (15, 5) code.

6.5   ROOTS AND PARITY CHECK MATRICES

Cyclic codes can of course also be represented in terms of their parity check or generator matrices. The parity check matrix $\mathbf{H}$ can be derived in a straightforward way from the roots of the generator. If a code polynomial $v(X)$ has a root β then

$$v(\beta) = 0$$

If v_n is the coefficient of X^n then

$$v_{n-1} \beta^{n-1} + \cdots + v_2 \beta^2 + v_1 \beta + v_0 = 0$$

or in vector form

$$\mathbf{v} \begin{bmatrix} \beta^{n-1} \\ \cdot \\ \cdot \\ \cdot \\ \beta^2 \\ \beta \\ 1 \end{bmatrix} = 0 \tag{6.1}$$

Similarly if there are j roots β_1 to β_j then

$$\mathbf{v} \begin{bmatrix} \beta_1^{n-1} & \beta_2^{n-1} & \cdot & \cdot & \beta_{j-1}^{n-1} & \beta_j^{n-1} \\ \cdot & \cdot & \cdot & \cdot & \cdot & \cdot \\ \cdot & \cdot & \cdot & \cdot & \cdot & \cdot \\ \beta_1^2 & \beta_2^2 & \cdot & \cdot & \beta_{j-1}^2 & \beta_j^2 \\ \beta_1 & \beta_2 & \cdot & \cdot & \beta_{j-1} & \beta_j \\ 1 & 1 & \cdot & \cdot & 1 & 1 \end{bmatrix} = 0 \tag{6.2}$$

but

$$\mathbf{v}\, \mathbf{H}^T = \mathbf{0}$$

which means that the large matrix in Equation (6.2), when transposed, will give the parity check matrix of the code.

The roots are polynomials in α and so may be regarded as vectors which themselves need to be transposed. Therefore

$$\mathbf{H} = \begin{bmatrix} \beta_1^{n-1^T} & \cdot & \cdot & \beta_1^{2^T} & \beta_1^{T} & 1^T \\ \beta_2^{n-1^T} & \cdot & \cdot & \beta_2^{2^T} & \beta_2^{T} & 1^T \\ \cdot & \cdot & \cdot & \cdot & \cdot & \cdot \\ \cdot & \cdot & \cdot & \cdot & \cdot & \cdot \\ \beta_{j-1}^{n-1^T} & \cdot & \cdot & \beta_{j-1}^{2^T} & \beta_{j-1}^{T} & 1^T \\ \beta_j^{n-1^T} & \cdot & \cdot & \beta_j^{2^T} & \beta_j^{T} & 1^T \end{bmatrix} \tag{6.3}$$

Only one of the roots β, β^2, β^4, β^8, etc. needs to be included in the parity check matrix as the inclusion of any one implies all the others.

Hamming codes

Hamming codes have generator polynomials which are primitive. Hence any primitive element can be a root of the code. If we take the element α as the root then

$$\mathbf{H} = \begin{bmatrix} \alpha^{n-1^T} & \cdot & \cdot & \alpha^{2^T} & \alpha^T & 1^T \end{bmatrix} \tag{6.4}$$

The powers of α are just all the nonzero elements of the field which leads to the conclusion that the columns of the parity check matrix contain all the possible combinations of 1 and 0. For example taking the code based on GF(8), for which $\alpha^3 + \alpha + 1 = 0$, gives

$$H = \begin{bmatrix} 1 & 1 & 1 & 0 & 1 & 0 & 0 \\ 0 & 1 & 1 & 1 & 0 & 1 & 0 \\ 1 & 1 & 0 & 1 & 0 & 0 & 1 \end{bmatrix}$$

This is in fact the parity check matrix for the cyclic Hamming code in Chapter 4.

Binary BCH codes

A double-error correcting binary BCH code might have roots of α, α^2, α^3 and α^4. Of these, only α and α^3 are independent, the others being implied by α, and so the parity check matrix is

$$H = \begin{bmatrix} \alpha^{n-1^T} & \cdot & \cdot & \alpha^{2^T} & \alpha^T & 1^T \\ \alpha^{3(n-1)^T} & & \cdot & \alpha^{3 \times 2^T} & \alpha^{3^T} & 1^T \end{bmatrix} \tag{6.5}$$

For the length 15 code, we obtain

$$H = \begin{bmatrix} \alpha^{14^T} & \cdot & \cdot & \alpha^{2^T} & \alpha^T & 1^T \\ \alpha^{12^T} & \cdot & \cdot & \alpha^{3 \times 2^T} & \alpha^{3^T} & 1^T \end{bmatrix}$$

or

$$H = \begin{bmatrix} 1 & 1 & 1 & 1 & 0 & 1 & 0 & 1 & 1 & 0 & 0 & 1 & 0 & 0 & 0 \\ 0 & 1 & 1 & 1 & 1 & 0 & 1 & 0 & 1 & 1 & 0 & 0 & 1 & 0 & 0 \\ 0 & 0 & 1 & 1 & 1 & 1 & 0 & 1 & 0 & 1 & 1 & 0 & 0 & 1 & 0 \\ 1 & 1 & 1 & 0 & 1 & 0 & 1 & 1 & 0 & 0 & 1 & 0 & 0 & 0 & 1 \\ 1 & 1 & 1 & 1 & 0 & 1 & 1 & 1 & 1 & 0 & 1 & 1 & 1 & 1 & 0 \\ 1 & 0 & 1 & 0 & 0 & 1 & 0 & 1 & 0 & 0 & 1 & 0 & 1 & 0 & 0 \\ 1 & 1 & 0 & 0 & 0 & 1 & 1 & 0 & 0 & 0 & 1 & 1 & 0 & 0 & 0 \\ 1 & 0 & 0 & 0 & 1 & 1 & 0 & 0 & 0 & 1 & 1 & 0 & 0 & 0 & 1 \end{bmatrix}$$

Note that this does not correspond to the usual definition of the parity check matrix for a systematic code. It will, however, be seen to be useful for decoding.

A triple-error correcting code would introduce an additional root α^5. The parity check matrix becomes

$$H = \begin{bmatrix} \alpha^{n-1^T} & \cdot & \cdot & \alpha^{2^T} & \alpha^T & 1^T \\ \alpha^{3(n-1)^T} & \cdot & \cdot & \alpha^{3 \times 2^T} & \alpha^{3^T} & 1^T \\ \alpha^{5(n-1)^T} & \cdot & \cdot & \alpha^{5 \times 2^T} & \alpha^{5^T} & 1^T \end{bmatrix} \tag{6.6}$$

Corresponding to this, the parity check matrix for the length 15 code is

$$H = \begin{bmatrix} \alpha^{14^T} & \cdot & \cdot & \alpha^{2^T} & \alpha^T & 1^T \\ \alpha^{12^T} & \cdot & \cdot & \alpha^{6^T} & \alpha^{3^T} & 1^T \\ \alpha^{10^T} & \cdot & \cdot & \alpha^{10^T} & \alpha^{5^T} & 1^T \end{bmatrix}$$

which in binary terms is

$$H = \begin{bmatrix} 1 & 1 & 1 & 1 & 0 & 1 & 0 & 1 & 1 & 0 & 0 & 1 & 0 & 0 & 0 \\ 0 & 1 & 1 & 1 & 1 & 0 & 1 & 0 & 1 & 1 & 0 & 0 & 1 & 0 & 0 \\ 0 & 0 & 1 & 1 & 1 & 1 & 0 & 1 & 0 & 1 & 1 & 0 & 0 & 1 & 0 \\ 1 & 1 & 1 & 0 & 1 & 0 & 1 & 1 & 0 & 0 & 1 & 0 & 0 & 0 & 1 \\ 1 & 1 & 1 & 1 & 0 & 1 & 1 & 1 & 1 & 0 & 1 & 1 & 1 & 1 & 0 \\ 1 & 0 & 1 & 0 & 0 & 1 & 0 & 1 & 0 & 0 & 1 & 0 & 1 & 1 & 0 \\ 1 & 1 & 0 & 0 & 0 & 1 & 1 & 0 & 0 & 0 & 1 & 1 & 0 & 0 & 0 \\ 1 & 0 & 0 & 0 & 1 & 1 & 0 & 0 & 0 & 1 & 1 & 0 & 0 & 0 & 1 \\ 0 & 0 & 0 & 0 & 0 & 0 & 0 & 0 & 0 & 0 & 0 & 0 & 0 & 0 & 0 \\ 1 & 1 & 0 & 1 & 1 & 0 & 1 & 1 & 0 & 1 & 1 & 0 & 1 & 1 & 0 \\ 1 & 1 & 0 & 1 & 1 & 0 & 1 & 1 & 0 & 1 & 1 & 0 & 1 & 1 & 0 \\ 1 & 0 & 1 & 1 & 0 & 1 & 1 & 0 & 1 & 1 & 0 & 1 & 1 & 0 & 1 \end{bmatrix}$$

Note, however, that there is one all-zero row and one that duplicates the previous one. Both can therefore be removed from the parity check matrix to give

$$H = \begin{bmatrix} 1 & 1 & 1 & 1 & 0 & 1 & 0 & 1 & 1 & 0 & 0 & 1 & 0 & 0 & 0 \\ 0 & 1 & 1 & 1 & 1 & 0 & 1 & 0 & 1 & 1 & 0 & 0 & 1 & 0 & 0 \\ 0 & 0 & 1 & 1 & 1 & 1 & 0 & 1 & 0 & 1 & 1 & 0 & 0 & 1 & 0 \\ 1 & 1 & 1 & 0 & 1 & 0 & 1 & 1 & 0 & 0 & 1 & 0 & 0 & 0 & 1 \\ 1 & 1 & 1 & 1 & 0 & 1 & 1 & 1 & 1 & 0 & 1 & 1 & 1 & 1 & 0 \\ 1 & 0 & 1 & 0 & 0 & 1 & 0 & 1 & 0 & 0 & 1 & 0 & 1 & 0 & 0 \\ 1 & 1 & 0 & 0 & 0 & 1 & 1 & 0 & 0 & 0 & 1 & 1 & 0 & 0 & 0 \\ 1 & 0 & 0 & 0 & 1 & 1 & 0 & 0 & 0 & 1 & 1 & 0 & 0 & 0 & 1 \\ 1 & 1 & 0 & 1 & 1 & 0 & 1 & 1 & 0 & 1 & 1 & 0 & 1 & 1 & 0 \\ 1 & 0 & 1 & 1 & 0 & 1 & 1 & 0 & 1 & 1 & 0 & 1 & 1 & 0 & 1 \end{bmatrix}$$

Again this is not the usual systematic form.

6.6 ALGEBRAIC DECODING

BCH codes allow an algebraic method of decoding. Consider the case of a double-error correcting code where there are errors at positions i and j. The syndrome is

$$s = e \, H^T$$

The syndrome has two components s_1 and s_3:

$$\begin{aligned} s_1 &= \alpha^i + \alpha^j \\ s_3 &= \alpha^{3j} + \alpha^{3j} \end{aligned} \tag{6.7}$$

Substituting the first into the second gives

$$s_1^2 \alpha^i + s_1 \alpha^{2i} + s_1^3 + s_3 = 0 \tag{6.8}$$

Any value of α^i which is a root of this equation will locate an error and, as the assignment of the parameters i and j is arbitrary, both error locations can be found from the same equation. Roots can be found by trying all possible values, which is better than having to try all possible combinations of positions, or by other techniques. This method is known as a *Chien search*.

Example

For the double-error correcting BCH code of length 15, if the received sequence is 101010110010101, the syndrome is 10010110. Thus

$$s_1 = \alpha^{14}$$

$$s_3 = \alpha^5$$

Substituting in Equation (6.8) gives

$$\alpha^{13+i} + \alpha^{14+2i} + \alpha^{12} + \alpha^5 = 0$$

The value $i = 2$ gives

$$\alpha^0 + \alpha^3 + \alpha^{12} + \alpha^5 = 0$$

The value $i = 13$ gives

$$\alpha^{11} + \alpha^{10} + \alpha^{12} + \alpha^5 = 0$$

The errors are therefore at positions 2 and 13 giving a transmitted sequence 111010110010001. The syndrome of this sequence is zero, showing that it is a codeword.

Of course it is possible that a single error may occur in a double-error correcting code. In that case Equation (6.7) becomes

$$s_1 = \alpha^i$$

$$s_3 = \alpha^{3i}$$

This condition can be recognized from the fact that $s_3 = (s_1)^3$ and the position of the error found directly from s_1.

6.7   BCH DECODING AND THE BCH BOUND

We have previously seen that binary BCH codes can be decoded by a polynomial with roots that indicate the positions of the errors. The method as described is suitable for single or double errors and can be extended to more errors, but with increasing complexity. To find a more general method, we note that the syndromes computed above are in fact the Fourier transform components in the positions of the frequency domain zeros, and we look for a frequency domain description of algebraic decoding.

We shall see in the next section that frequency domain decoding uses a polynomial whose zero coefficients indicate the locations of errors in the time domain and whose roots in the frequency domain can therefore be used to locate the errors. For the moment we will use such a polynomial to prove a property of BCH codes, namely that $2t$ consecutive zero-valued spectral components are sufficient to guarantee t-error correction.

Suppose we have a code vector $c(X)$ with fewer than d nonzero components, and its spectrum $C(z)$ has $d - 1$ consecutive zeros. We define a polynomial $\lambda(X)$ such that it is zero where $c(X)$ is nonzero and let the positions of the zeros be denoted i_j. The polynomial $\lambda(X)$ is usually called the *error locator polynomial* because, as we shall see, there is no codeword apart from the all-zero sequence which satisfies the defined conditions.

The zero components in $\lambda(X)$ mean that each α^{-i_j} will be a root of the transform $\Lambda(z)$ of $\lambda(X)$, or

$$\Lambda(z) = \prod_{j=1}^{v}\left(1 + z^{i_j}\right)$$

Note from the above definition of $\Lambda(z)$ that Λ_0 is equal to 1. The polynomial $\Lambda(z)$ is known as the *connection polynomial*.

Now in the time domain, $\lambda_i c_i = 0$ for all i, therefore in the frequency domain, replacing multiplication with a convolution:

$$\sum_{j=0}^{n-1} \Lambda_j\, C_{k-j} = 0$$

The degree v of $\Lambda(z)$ is at most $d - 1$ and $\Lambda_0 = 1$, which leaves us with

$$C_k = \sum_{j=1}^{d-1} \Lambda_j\, C_{k-j} \qquad (6.9)$$

This is the equation for the output of a linear feedback shift register with feedback polynomial $\Lambda(z)$, as shown in Figure 6.1.

If we know any $d - 1$ consecutive values of C_j we can use the shift registers with feedback to generate all the rest. We know, however that there are $d - 1$ consecutive

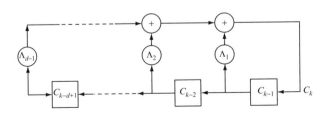

Figure 6.1 Shift register representation of connection polynomial

zeros and using them to initialize the feedback shift register will generate a string of zeros. Thus C_j must be zero for all j and all terms in $c(X)$ are zero. This proves that if there are $2t$ consecutive zeros in the spectrum, the nonzero codewords must have weight at least $2t + 1$ and the code can therefore correct at least t errors.

Although we are left with the possibility that the code may be able to correct more than t errors, the BCH decoding method will use the $2t$ consecutive zeros in a way that will correct up to t errors, and any extra capabilities of the code will go into error detection.

6.8   DECODING IN THE FREQUENCY DOMAIN

Assume that we transmit a code sequence $c(X)$ and that it is received with the addition of an error sequence $e(X)$. We take the transform of the received sequence, giving

$$R(z) = C(z) + E(z)$$

where $R(z)$, $C(z)$ and $E(z)$ are the transforms of the received sequence, the codeword and the error sequence respectively. We know that $C(z)$ is zero in $2t$ consecutive spectral locations, so we can use these locations to give us a window on $E(z)$, i.e. $2t$ components of $E(z)$ can easily be obtained and can be considered to form a syndrome $S(z)$.

We assume that there are $v \leq t$ errors and define an error locator polynomial $\lambda(X)$ such that it is zero in the positions where $e(X)$ is nonzero. The product of the received sequence and the error locator sequence in the time domain will therefore be zero, which means that in the frequency domain the convolution will be zero

$$\sum_{j=0}^{t} \Lambda_j E_{k-j} = 0 \tag{6.10}$$

Here we have used the fact that $\lambda(X)$ has at most t zeros; hence, $\Lambda(z)$ has at most t roots and is therefore a polynomial of degree no greater than t. If we know the error spectrum from positions m to $m + 2t - 1$, we can form t equations by letting k take values from $m + t$ to $m + 2t - 1$. Assuming the spectral zeros of the code are from positions 1 to $2t$, the equations are as follows:

$$\Lambda_0 E_{t+1} + \Lambda_1 E_t + \cdots + \Lambda_t E_1 = 0$$
$$\cdots$$
$$\cdots \tag{6.11}$$
$$\Lambda_0 E_{2t-1} + \Lambda_1 E_{2t-2} + \cdots + \Lambda_t E_{t-1} = 0$$
$$\Lambda_0 E_{2t} + \Lambda_1 E_{2t-1} + \cdots + \Lambda_t E_t = 0$$

This set of t equations in $t + 1$ unknowns is called the *key equation*, which we can solve for the different values of Λ_j provided we impose an arbitrary value on one of the roots, corresponding to the fact that the value of the error locator polynomial is arbitrary in the nonzero positions in the time domain. In practice the condition we impose is $\Lambda_0 = 1$. The procedure for solving the key equation may be straightforward if t is small, but in the general case special methods have to be devised which are computationally efficient and which take into account the fact that the number of errors may be less than t. One such method, Euclid's algorithm, will be explained in a later section. For the moment it is sufficient to believe that the task can be accomplished.

6.9 DECODING EXAMPLES FOR BINARY BCH CODES

Take the example of the (15, 7) binary BCH code from Section 6.6 where we received a sequence 101010110010101 which was found to have two errors, in positions 2 and 13. To find the syndrome, we calculate the Fourier transform of the received sequence in the positions of the zeros, i.e. positions 1, 2, 3 and 4. In fact the value in position 2 is, through conjugacy, the square of the value in position 1 and the value in position 4 the square of that in position 2. We therefore need calculate only the values in positions 1 and 3 (found previously to be (α^{14} and α^5) before filling in the values for positions 2 and 4. The syndrome polynomial is therefore

$$S(z) = \alpha^{11}z^4 + \alpha^5 z^3 + \alpha^{13}z^2 + \alpha^{14}z$$

The key equation becomes

$$\alpha^5 + \alpha^{13}\Lambda_1 + \alpha^{14}\Lambda_2 = 0$$

$$\alpha^{11} + \alpha^5 \Lambda_1 + \alpha^{13}\Lambda_2 = 0$$

To solve, we can multiply each term in the second equation by α and add to the first equation to eliminate Λ_2". This gives

$$\Lambda_1 = \alpha^{14}$$

Hence

$$\Lambda_2 = 1$$

Decoding can now be carried out by finding the roots of $z^2 + \alpha^{14}z + 1$, which are α^2 and α^{13}. Hence the errors are at positions -2 and -13, or 13 and 2, respectively.

Suppose now that only one error occurred, say the error in position 13. The syndromes are found to be $S_1 = \alpha^{13}$, $S_3 = \alpha^9$, giving a full syndrome polynomial of

$$S(z) = \alpha^7 z^4 + \alpha^9 z^3 + \alpha^{11}z^2 + \alpha^{13}z$$

The key equation becomes

$$\alpha^9 + \alpha^{11}\Lambda_1 + \alpha^{13}\Lambda_2 = 0$$

$$\alpha^7 + \alpha^9 \Lambda_1 + \alpha^{11}\Lambda_2 = 0$$

It can be seen that these two equations are not linearly independent, the first being merely α^2 times the second. Only one unknown can therefore be found and the connection polynomial must be of degree 1, i.e. $\Lambda_2 = 0$. Inserting this condition gives $\Lambda_1 = \alpha^{-2} = \alpha^{13}$. The connection polynomial is therefore $\alpha^{13}z + 1$, which has root α^{-13}, indicating an error at position 13. Any solution method therefore needs to take account of the possibility that there may be fewer errors than the maximum corrected by the code. Of course, in this case the occurrence of a single error can be recognized through the condition $s_3 = (s_1)^3$ from Section 6.6 and its location determined from the value of s_1.

Another possibility is that too many errors may occur and in that case the decoder may give the wrong result or may detect the errors but be unable to correct them. We need to know, however, how the decoding failure will arise.

Suppose we insert a third error at position 10 into our received sequence. The syndromes are found to be $s_1 = \alpha^{11}$, $s_3 = \alpha^{10}$. The full syndrome polynomial is

$$S(z) = \alpha^{14}z^4 + \alpha^{10}z^3 + \alpha^7 z^2 + \alpha^{11}z$$

The key equation is

$$\alpha^{10} + \alpha^7\Lambda_1 + \alpha^{11}\Lambda_2 = 0$$
$$\alpha^{14} + \alpha^{10}\Lambda_1 + \alpha^7\Lambda_2 = 0$$

Eliminating Λ_2 gives

$$\alpha^{12} + \alpha^1\Lambda_1 = 0$$
$$\Lambda_1 = \alpha^{11}, \; \Lambda_2 = \alpha^1$$

The connection polynomial is therefore $\alpha^1 z^2 + \alpha^{11}z + 1$. We now try to find the roots as shown in Table 6.2. It is seen that no roots are found. The general condition for detected uncorrectable errors is that the number of roots found is less than the degree of the connection polynomial.

Table 6.2 Chien search for triple-error example

Root	Evaluation	Root	Evaluation
α^0	$\alpha^1\alpha^0 + \alpha^{11}\alpha^0 + \alpha^0 = \alpha^{13}$	α^8	$\alpha^1\alpha^1 + \alpha^{11}\alpha^8 + \alpha^0 = \alpha^5$
α^1	$\alpha^1\alpha^2 + \alpha^{11}\alpha^1 + \alpha^0 = \alpha^5$	α^9	$\alpha^1\alpha^3 + \alpha^{11}\alpha^9 + \alpha^0 = \alpha^2$
α^2	$\alpha^1\alpha^4 + \alpha^{11}\alpha^2 + \alpha^0 = \alpha^9$	α^{10}	$\alpha^1\alpha^5 + \alpha^{11}\alpha^{10} + \alpha^0 = \alpha^0$
α^3	$\alpha^1\alpha^6 + \alpha^{11}\alpha^3 + \alpha^0 = \alpha^4$	α^{11}	$\alpha^1\alpha^7 + \alpha^{11}\alpha^{11} + \alpha^0 = \alpha^{12}$
α^4	$\alpha^1\alpha^8 + \alpha^{11}\alpha^4 + \alpha^0 = \alpha^9$	α^{12}	$\alpha^1\alpha^9 + \alpha^{11}\alpha^{12} + \alpha^0 = \alpha^4$
α^5	$\alpha^1\alpha^{10} + \alpha^{11}\alpha^5 + \alpha^0 = \alpha^{13}$	α^{13}	$\alpha^1\alpha^{11} + \alpha^{11}\alpha^{13} + \alpha^0 = \alpha^2$
α^6	$\alpha^1\alpha^{12} + \alpha^{11}\alpha^6 + \alpha^0 = \alpha^3$	α^{14}	$\alpha^1\alpha^{13} + \alpha^{11}\alpha^{14} + \alpha^0 = \alpha^{12}$
α^7	$\alpha^1\alpha^{14} + \alpha^{11}\alpha^7 + \alpha^0 = \alpha^3$		

6.10 POLYNOMIAL FORM OF THE KEY EQUATION

In the previous section we have been involved in solving simultaneous equations in order to carry out error correction. This we can do manually by substitution or other means. For automatic implementation we need to find an approach which can be efficiently and routinely implemented. To do this we first of all convert the key equation from its expression as a summation into a polynomial format.

We wish to find a solution to the key equation, which can be expressed as

$$\sum_{j=0}^{t} \Lambda_j E_{k-j} = 0 \qquad t \le k \le 2t - 1$$

This expression represents a convolution that can, alternatively, be given a polynomial form. We have seen on a number of occasions previously that the convolution of two sequences corresponds to the product of the sequence polynomials. In this case it is equivalent to all the terms of degree between t and $2t - 1$ in $\Lambda(z)E(z)$. Hence we can say

$$\Lambda(z)E(z) = f(z)z^{2t} + \Omega(z) \tag{6.12}$$

The terms in the left-hand side of degree t to $2t - 1$ are zero. The terms of degree $2t$ or more are represented by $f(z)z^{2t}$. Here, $\Omega(z)$ represents the terms of degree less than t. It is known as the error evaluator polynomial because it is used for such a purpose in the decoding of multilevel BCH codes such as Reed Solomon codes.

There are two commonly used methods for solving the key equation in this form. They are Euclid's algorithm and the Berlekamp–Massey algorithm. The latter is more efficient, but is more difficult to understand than Euclid's method and is related to it. We shall study the Euclid algorithm in detail and merely outline the steps involved in the other.

6.11 EUCLID'S METHOD

The Euclid algorithm is most commonly encountered in finding the lowest common multiples of numbers. In the process of so doing it identifies common factors so that these can be taken into account in computing the lowest common multiple. Many authors preface a discussion of Euclid's method with an illustration of its use for such a purpose. Unfortunately, the connection between using it in this way and solving the key equation is not easy to spot and the numeric application may not be of much help in understanding the application to polynomials. For that reason, I prefer to confine the discussion to the solution of equations involving polynomials. Euclid's method enables us to find minimum degree solutions for polynomials $f(z)$ and $g(z)$ such that

$$a(z)f(z) + b(z)g(z) = r(z)$$

where $r(z)$ is known to have degree less than some fixed value. In our case, $r(z)$ will have degree $< t$, $a(z) = z^{2t}$ and $b(z)$ is the syndrome polynomial $S(z)$. The polynomial $g(z)$ will give us $\Lambda(z)$, which is what we need to know. The method involves repeated division of polynomials until a remainder of degree $< t$ is found.

The first step is to divide $a(z)$ by $b(z)$ to find the quotient $q_1(z)$ and remainder $r_1(z)$:

$$a(z) = q_1(z) b(z) + r_1(z) \tag{6.13}$$

If the degree of $r_1(z)$ is less than t then we have reached our solution with $f(z) = 1$, $g(z) = q_1(z)$ and $r(z) = r_1(z)$. Otherwise set $g_1(z) = q_1(z)$ and proceed to the next stage. The second step is to divide $b(z)$ by $r_1(z)$ giving

$$b(z) = q_2(z)r_1(z) + r_2(z) \tag{6.14}$$

Note that the degree of $r_2(z)$ must be less than that of $r_1(z)$ so that this process is reducing the degree of the remainder. If we eliminate $r_1(z)$ from Equations (6.13) and (6.14) we obtain

$$q_2(z)a(z) = [q_2(z)g_1(z) + 1]b(z) + r_2(z) \tag{6.15}$$

Set $g_2(z) = q_2(z)g_1(z) + 1$. If the degree of $r_2(z)$ is less than t then $g(z) = g_2(z)$; otherwise, continue to the next step.

The third step continues in similar vein, dividing $r_1(z)$ by $r_2(z)$:

$$r_1(z) = q_2(z)r_2(z) + r_3(z) \tag{6.16}$$

Again the degree of the remainder is decreasing. Using Equations (6.14) and (6.15) to eliminate $r_1(z)$ and $r_2(z)$ gives

$$[1 + q_2(z)q_3(z)] a(z) = [g_1(z) + q_3(z)g_2(z)] b(z) + r_3(z) \tag{6.17}$$

If the degree of $r_3(z)$ is less than t then $g(z) = g_3(z) = q_3(z)g_2(z) + g_1(z)$.

The method continues in this way until a remainder of degree less than t is found, at each stage setting

$$g_n(z) = q_n(z)g_{n-1}(z) + g_{n-2}(z) \qquad [g_0(z) = 1, \ g_{-1}(z) = 0] \tag{6.18}$$

Summary of Euclid's algorithm

Set $g_{-1}(z) = 0$, $g_0(z) = 1$.

Set $n = 1$. Divide z^{2t} by $S(z)$ to find quotient $q_1(z)$ and remainder $r_1(z)$. Calculate $g_1(z)$ from Equation (6.18).

While degree of remainder is greater than or equal to t, continue by incrementing n, divide previous divisor by previous remainder and calculate $g_n(z)$ by Equation (6.18).

When desired degree of remainder is obtained, set $\Lambda(z) = g_n(z)$.

Example

We shall now solve the key equation for the double-error example from Section 6.9, using the Euclid algorithm. We treat the syndrome with 4 terms as a degree-3 polynomial

$$S(z) = \alpha^{11} z^3 + \alpha^5 z^2 + \alpha^{13} z + \alpha^{14}$$

Divide z^4 by $S(z)$ to give

$$z^4 = (\alpha^4 z + \alpha^{13})(\alpha^{11} z^3 + \alpha^5 z^2 + \alpha^{13} z + \alpha^{14}) + \alpha^6 z^2 + \alpha^5 z + \alpha^{12}$$

Set $g_1(z) = \alpha^4 z + \alpha^{13}$
Divide $S(z)$ by $\alpha^6 z^2 + \alpha^5 z + \alpha^{12}$ to give

$$\alpha^{11} z^3 + \alpha^5 z^2 + \alpha^{13} z + \alpha^{14} = (\alpha^5 z + \alpha^9)(\alpha^6 z^2 + \alpha^5 z + \alpha^{12}) + \alpha^8$$

Set $g_2(z) = (\alpha^5 z + \alpha^9)(\alpha^4 z + \alpha^{13}) + 1 = \alpha^9 z^2 + \alpha^8 z + \alpha^9$.

As the remainder is of degree < 2, this is the end of Euclid's algorithm. The result for $g_2(z)$ is the connection polynomial $\Lambda(z)$. It can be seen to be a factor of α^9 times the result obtained in Section 6.9 and hence will have the same roots.

Applying Euclid's algorithm to a single-error example would terminate with a polynomial of degree 1. This is left as an exercise for the reader.

6.12   BERLEKAMP–MASSEY ALGORITHM

Another way to find the connection polynomial is to use the Berlekamp–Massey algorithm. This algorithm is difficult to understand, although it synthesizes directly the shift registers with feedback shown in Figure 6.1. It is also simple to implement; consequently it will be described here but an explanation of why it works will not be attempted. In the following description, the parameter l represents the degree of the error locator polynomial and n represents the degree of the syndrome polynomial being examined.

Table 6.3 shows a Pascal description of the algorithm in the left column and, in the other columns, the steps in the calculations for our double-error BCH example. As before, the syndrome will be treated as a polynomial of degree 3.

The algorithm terminates with $\Lambda(z)$ holding the correct coefficients of the feedback polynomial. Applied to a single-error example it will also terminate correctly with a polynomial of degree 1. Note that on every second iteration of the main loop the value of δ is zero. This is because every second syndrome component in a binary code is predictable from previous values. The algorithm could therefore be simplified for the binary case to multiply $D(z)$ by z^2 at each iteration and to increment n by 2.

This version of the Berlekamp–Massey algorithm involves a division and, as mentioned in Chapter 5, finding the multiplicative inverse of an element is not straightforward. Given that only the roots of the connection polynomial are important,

Table 6.3   Berkekamp–Massey algorithm for binary BCH example

begin				
$l := 0$:	0			
$n := 0$;	0			
$k := -1$;	-1			
$\Lambda(z) := 1$;	1			
$D(z) := z\Lambda(z)$;	z			
while $(n < 2t)$ do	$0 < 4$	$1 < 4$	$2 < 4$	$3 < 4$
begin				
$\delta := \sum\limits_{i=0}^{l} \Lambda_i S_{n-i}$;	α^{14}	0	α^{14}	0
if $(\delta \neq 0)$ then	$\alpha^{14} \neq 0$	$0 = 0$	$\alpha^{14} \neq 0$	$0 = 0$
begin				
$\Lambda^*(z) = \Lambda(z) + \delta \times D(z)$;	$\alpha^{14}z + 1$		$z^2 + \alpha^{14}z + 1$	
if $(l < n - k)$	$0 < 1$		$1 < 2$	
begin				
$l^* := n - k$;	1		2	
$k := n - l$;	0		1	
$D(z) := \Lambda(z)/\delta$;	α^1		$z + \alpha$	
$l := l^*$	1		2	
end;				
$\Lambda(z) := \Lambda^*(z)$	$\alpha^{14}z + 1$		$z^2 + \alpha^{14}z + 1$	
end;				
$D(z) := zD(z)$;	$\alpha^1 z$	$\alpha^1 z^2$	$z^2 + \alpha z$	$z^3 + \alpha z^2$
$n := n + 1$	1	2	3	4
end;				
end.				

we could multiply it by any constant. Thus we could eliminate the division in the line $D(z) := \Lambda(z)/\delta$ provided that on the line $\Lambda^*(z) = \Lambda(z) + \delta \times D(z)$ we first multiply the value of $\Lambda(z)$ by the previously applied version of δ. We therefore introduce a variable δ^* and amend the algorithm to that shown in Table 6.4.

The working shown in Table 6.4 is again the stages in calculating the connection polynomial for the double-error example of Section 6.9. The connection polynomial is $\alpha^{14}z^2 + \alpha^{13}z + \alpha^{14}$ which is just α^{14} times the previous result and therefore has the same roots.

6.13   CONCLUSION

As indicated at the beginning of this chapter, BCH codes have a long history and are therefore well represented in standard text books [5–9]. The inversionless Berlekamp–Massey algorithm given here is similar to that published in [10] and another implementation is embodied in [11]. The performance and applicability of binary BCH codes are discussed in Chapter 8.

Table 6.4 Inversionless version of Berlekamp–Massey algorithm

begin				
$l := 0;$	0			
$n := 0;$	0			
$k := -1;$	-1			
$\Lambda(z) := 1;$	1			
$D(z) := z\Lambda(z);$	z			
$\delta^* := 1;$	1			
while $(n < 2t)$ do	$0 < 4$	$1 < 4$	$2 < 4$	$3 < 4$
begin				
$\displaystyle\delta := \sum_{i=0}^{l} \Lambda_i S_{n-i};$	α^{14}	0	α^{14}	0
if $(\delta \neq 0)$ then	$\alpha^{14} \neq 0$	$0 = 0$	$\alpha^{14} \neq 0$	$0 = 0$
begin				
$\Lambda^*(z) = \delta^* \times \Lambda(z) + \delta \times D(z);$	$\alpha^{14}z + 1$		$\alpha^{14}z^2 + \alpha^{13}z + \alpha^{14}$	
if $(l < n - k)$	$0 < 1$		$1 < 2$	
begin				
$l^* := n - k;$	1		2	
$k := n - l;$	0		1	
$\delta^* := \delta;$	α^{14}		α^{14}	
$D(z) := \Lambda(z);$	α^0		$\alpha^{14}z + 1$	
$l := l^*$	1		2	
end;				
$\Lambda(z) := \Lambda^*(z)$	$\alpha^{14}z + 1$		$\alpha^{14}z^2 + \alpha^{13}z + \alpha^{14}$	
end;				
$D(z) := zD(z);$	$\alpha^0 z$	$\alpha^0 z^2$	$\alpha^{14}z^2 + z$	$\alpha^{14}z^3 + z^2$
$n := n + 1$	1	2	3	4
end;				
end.				

6.14 EXERCISES

1 Given that

$$X^{15} + 1 = (X + 1)(X^4 + X + 1)(X^4 + X^3 + 1)(X^4 + X^3 + X^2 + X + 1)$$
$$(X^2 + X + 1)$$

find the generator polynomial of a triple-error correcting BCH code of length 15, assuming that GF(16) is created using the primitive polynomial $X^4 + X^3 + 1$.

2 For the double-error correcting BCH code defined in Section 6.5, decode the sequence 100010110010001.

3 Describe the frequency domain representation (i.e. positions of the zeros) of a triple-error correcting BCH code of length 31. Why would you not put a zero in position zero of the frequency domain?

4 Find the generator polynomial of a triple-error correcting BCH code of length 15, assuming that GF(16) is created using the primitive polynomial $X^4 + X^3 + 1$.

5 For the BCH code of length 15 with roots , α, α^2, α^3, and α^4 in GF(16) (created using the primitive polynomial $X^4 + X^3 + 1$), find the generator polynomial and decode the following sequences

100010110010001
101110100101001
100110000011111

6.15 REFERENCES

1 A. Hocquenghem, *Codes correcteurs d'erreurs*, Chiffres, Vol. 2, pp. 147–156, 1959.
2 R.C. Bose and D.K. Ray-Chaudhuri, *On a class of error-correcting binary group codes*, Information and Control, Vol. 3, pp. 68–79, 1960.
3 I.S. Reed and G. Solomon, *Polynomial codes over certain finite fields*, J. Soc. Indust. Applied Math. Vol. 8, pp. 300–304, 1960.
4 D.C. Gorenstein and N. Zierler, *A class of error-correcting codes in p^m symbols*, J. Soc. Indust. Applied Math. Vol. 9, pp. 207–214, 1961.
5 R.E. Blahut, *Theory and Practice of Error Control Codes*, Addison Wesley, 1983.
6 G.C. Clark and J.B. Cain, *Error-Correction Coding for Digital Communications*, Plenum Press, 1981.
7 A.M. Michelson and A.H. Levesque, *Error-Control Techniques for Digital Communication*, John Wiley & Sons, 1985.
8 S.B. Wicker, *Error Control Systems for Digital Communication and Storage*, Prentice Hall, 1994.
9 M. Bossert, *Channel Coding for Telecommunications*, John Wiley and Sons, 1999.
10 Y. Xu, *Implementation of Berlekamp–Massey Algorithm without inversion*, IEE Proceedings-I, Vol. 138, No. 3, pp. 138–140, June 1991.
11 H-C. Chang and C-S.B. Shung, *Method and apparatus for solving key equation polynomials in decoding error correction codes*, US Patent US6119262, September 2000.

7

Reed Solomon codes

7.1   INTRODUCTION

It was pointed out in Chapter 6 that Reed Solomon codes are a special example of multilevel BCH codes. Because the symbols are nonbinary, an understanding of finite field arithmetic is essential even for encoding. Moreover the decoding methods will be similar to those encountered for binary BCH codes, so some familiarity with Chapter 6 will also be advisable for the understanding of this chapter.

There are two distinctly different approaches to the encoding of Reed Solomon codes. One works in the time domain through calculation of parity check symbols. The other works in the frequency domain through an inverse Fourier transform. We shall meet the time domain technique first as it is more likely to be encountered in practice. It was pointed out in Chapter 6 that the standard decoding method is essentially a frequency domain technique, but we will see that a modification is needed for RS codes which can be achieved in two different ways. There is also a purely time domain decoding method for RS codes – the Welch–Berlekamp algorithm – which will be presented towards the end of the chapter.

7.2   GENERATOR POLYNOMIAL FOR A
REED SOLOMON CODE

A Reed Solomon code is a special case of a BCH code in which the length of the code is one less than the size of the field over which the symbols are defined. It consists of sequences of length $q - 1$ whose roots include $2t$ consecutive powers of the primitive element of $GF(q)$. Alternatively, the Fourier transform over $GF(q)$ will contain $2t$ consecutive zeros. Note that because both the roots and the symbols are specified in $GF(q)$, the generator polynomial will have only the specified roots; there will be no conjugates. Similarly the Fourier transform of the generator sequence will be zero in only the specified $2t$ consecutive positions.

A consequence of there being only $2t$ roots of the generator polynomial is that there are only $2t$ parity checks. This is the lowest possible value for any t-error correcting code and is known as the Singleton bound (see Chapter 8, Section 8.5).

To construct the generator for a Reed Solomon code, we need only to construct the appropriate finite field and choose the roots. Suppose we decide that the roots will be from α^i to α^{i+2t-1}, the generator polynomial will be

$$g(X) = (X + \alpha^i)(X + \alpha^{i+1}) \cdots (X + \alpha^{i+2t-2})(X + \alpha^{i+2t-1})$$

In contrast to the case with binary BCH codes, the choice of value of i will not affect the dimension or the minimum distance of the code because there are no conjugates to consider.

Example

Suppose we wish to construct a double-error correcting, length 7 RS code; we first construct GF(8) using the primitive polynomial $X^3 + X + 1$ as shown in Table 7.1. We decide to choose $i = 0$, placing the roots from α^0 to α^3. The generator polynomial is

$$g(X) = (X + \alpha^0)(X + \alpha^1)(X + \alpha^2)(X + \alpha^3)$$
$$g(X) = X^4 + (\alpha^0 + \alpha^1 + \alpha^2 + \alpha^3)X^3 + (\alpha^1 + \alpha^2 + \alpha^3 + \alpha^4 + \alpha^5)X^2$$
$$+ (\alpha^3 + \alpha^4 + \alpha^5 + \alpha^6)X + \alpha^6$$
$$g(X) = X^4 + \alpha^2 X^3 + \alpha^5 X^2 + \alpha^5 X + \alpha^6$$

Table 7.1   GF(8)

Element	Polynomial
0	000
α^0	001
α^1	010
α^2	100
α^3	011
α^4	110
α^5	111
α^6	101

7.3   TIME DOMAIN ENCODING FOR REED SOLOMON CODES

The encoding of a Reed Solomon code can be done by a long division method similar to that of Chapter 4 (Section 4.9) or, equivalently, by shift registers with feedback.

The long division method is slightly more complicated than in Chapter 4 because we need to subtract multiples of the divisor from the dividend so that the degree of the remainder is reduced. In the binary case the multiple is always 0 or 1, but here we need to choose an appropriate value from the finite field. Also the binary data will first need to be mapped onto finite field symbols and the result mapped back to binary values. An example will illustrate this.

Example

For the Reed Solomon code above, encode the data sequence 111001111. Assuming the polynomial mapping in Table 7.1, the data maps to symbols $\alpha^5\,\alpha^0\,\alpha^5$. Four zeros are appended, corresponding to the four parity checks to be generated, and the divisor is the generator sequence.

$$
\begin{array}{r}
\alpha^5\ \ 0\ \ \alpha^2 \\
\alpha^0\ \alpha^2\ \alpha^5\ \alpha^5\ \overline{\alpha^6)\ \alpha^5\ \ \alpha^0\ \ \alpha^5\ \ 0\ \ 0\ \ 0\ \ 0} \\
\underline{\alpha^5\ \ \alpha^0\ \ \alpha^3\ \ \alpha^3\ \alpha^4} \\
\alpha^2\ \ \alpha^3\ \alpha^4\ 0\ \ 0 \\
\underline{\alpha^2\ \ \alpha^4\ \alpha^0\ \alpha^0\ \alpha^1} \\
\alpha^6\ \ \alpha^5\ \alpha^0\ \alpha^1
\end{array}
$$

As shown, the multipliers used were $\alpha^5\ 0$ and α^2. The remainder is $\alpha^6\,\alpha^5\,\alpha^0\,\alpha^1$ so that the codeword is $\alpha^5\,\alpha^0\,\alpha^5\,\alpha^6\,\alpha^5\,\alpha^0\,\alpha^1$. Expressed as a binary sequence this is 111001111101111001010.

The encoder circuit for a Reed Solomon code is shown in Figure 7.1. This is almost identical to Figure 4.3 except that the value of g_0 is not necessarily 1 and there are multipliers in every feedback connection rather than just connection or no connection as there was for the binary case. In fact all the feedback terms will be nonzero for a Reed Solomon code.

The encoder for the above example is shown in Figure 7.2 and the stages in the encoding in Table 7.2. The example terminates with the registers containing the values $\alpha^6\,\alpha^5\,\alpha^0\,\alpha^1$, the parity checks for this codeword.

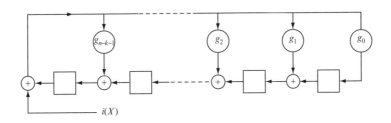

Figure 7.1 Schematic of Reed Solomon encoder

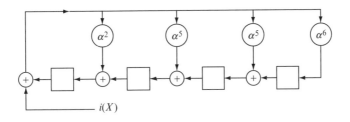

Figure 7.2 (7, 3) Reed Solomon encoder

ERROR CONTROL CODING

Table 7.2 Parity generation for (7, 3) Reed Solmon example

Input	Feedback	Register multiplier	X^3 α^2	X^2 α^5	X^1 α^5	X^0 α^6
			0	0	0	0
α^5	α^5		α^0	α^3	α^3	α^4
α^0	0		α^3	α^3	α^4	0
α^5	α^2		α^6	α^5	α^0	α^1

7.4 DECODING REED SOLOMON CODES

The frequency domain algebraic decoding method explained in Chapter 6 (Section 6.8) is used to decode Reed Solomon codes. There are, however, two differences. The first is that all the syndromes must be calculated from the received sequence; we cannot use conjugacy to find some of the values. Secondly, once we have located the errors we need to find their value to carry out the correction; this will require an extra stage in the decoding.

Error value calculation can be done using the *Forney algorithm*. Recall that in Chapter 6 (Equation (6.12)) we found that

$$\Lambda(z)\,E(z) = f(z)z^{2t} + \Omega(z)$$

where $\Lambda(z)$ is the connection polynomial, $E(z)$ can be taken as the frequency domain syndrome terms expressed as a polynomial and $\Omega(z)$ is a polynomial of degree $t - 1$ known as the error evaluator polynomial. We calculate the error evaluator polynomial and also $\Lambda'(z)$, the formal derivative of the connection polynomial. This is found to be

$$\begin{aligned}
&\text{(even } t) \quad \Lambda_{t-1}\,z^{t-2} + \Lambda_{t-3}\,z^{t-4} + \cdots + \Lambda_1 \\
&\text{(odd } t) \quad \Lambda_t\,z^{t-1} + \Lambda_{t-2}\,z^{t-3} + \cdots + \Lambda_1
\end{aligned} \tag{7.1}$$

In other words, get rid of the zero coefficient of Λ and then set all the odd terms in the resulting series to zero.

The error value in position m is now

$$e_m = \Omega(z)/z^{(1-i)}\Lambda'(z) \tag{7.2}$$

evaluated at $z = \alpha^{-m}$. The parameter i is the starting location of the roots of the generator polynomial.

7.5   REED SOLOMON DECODING EXAMPLE

Consider the codeword $\alpha^5\ \alpha^0\ \alpha^5\ \alpha^6\ \alpha^5\ \alpha^0\ \alpha^1$ previously generated for the double-error correcting (7, 3) RS code. We create errors in positions 5 and 3, assuming that we receive $\alpha^5\ \alpha^4\ \alpha^5\ \alpha^3\ \alpha^5\ \alpha^0\ \alpha^1$. The frequency domain syndrome of this sequence is

$$S_0 = \alpha^5 + \alpha^4 + \alpha^5 + \alpha^3 + \alpha^5 + \alpha^0 + \alpha^1 = \alpha^0$$
$$S_1 = \alpha^5 \cdot \alpha^6 + \alpha^4 \cdot \alpha^5 + \alpha^5 \cdot \alpha^4 + \alpha^3 \cdot \alpha^3 + \alpha^5 \cdot \alpha^2 + \alpha^0 \cdot \alpha^1 + \alpha^1 = \alpha^1$$
$$S_2 = \alpha^5 \cdot \alpha^5 + \alpha^4 \cdot \alpha^3 + \alpha^5 \cdot \alpha^1 + \alpha^3 \cdot \alpha^6 + \alpha^5 \cdot \alpha^4 + \alpha^0 \cdot \alpha^2 + \alpha^1 = \alpha^0$$
$$S_3 = \alpha^5 \cdot \alpha^4 + \alpha^4 \cdot \alpha^1 + \alpha^5 \cdot \alpha^5 + \alpha^3 \cdot \alpha^2 + \alpha^5 \cdot \alpha^6 + \alpha^0 \cdot \alpha^3 + \alpha^1 = 0$$

Now we form the key equation for which the solution is

$$\alpha^0 \Lambda_2 + \alpha^1 \Lambda_1 + \alpha^0 = 0$$
$$\alpha^1 \Lambda_2 + \alpha^0 \Lambda_1 = 0$$

$$\alpha^6 \Lambda_1 = \alpha^1$$
$$\Lambda_1 = \alpha^2 \quad \Lambda_2 = \alpha^1$$
$$\Lambda(z) = \alpha^1 z^2 + \alpha^2 z + 1$$

The roots of the connection polynomial are α^4 (as $\alpha^1 \alpha^8 + \alpha^2 \alpha^4 + 1 = 0$) and α^2 (as $\alpha^1 \alpha^4 + \alpha^2 \alpha^2 + 1 = 0$). Having found two roots for a connection polynomial of degree 2 indicates successful error correction with errors located at positions -4 and -2, i.e. positions 3 and 5. We now calculate the error evaluator polynomial, taking the powers from 0 to $t - 1$ of $S(z)\Lambda(z)$.

$$\Omega(z) = (S_0 \Lambda_1 + S_1 \Lambda_0)z + S_0 \Lambda_0$$
$$\Omega(z) = (\alpha^1 + \alpha^3 + \alpha^0)z^2 + (\alpha^2 + \alpha^1)z + \alpha^0$$
$$\Omega(z) = \alpha^4 z + \alpha^0$$

For the assumed $g(X)$ in which $i = 0$, we want $z\Lambda'(z)$ as the denominator which is just $\Lambda(z)$ with all the even coefficients set to zero. Therefore

$$e_m = \left.\frac{\Omega(z)}{z\Lambda'(z)}\right|_{z=\alpha^{-m}} = \left.\frac{\alpha^4 z + \alpha^0}{\alpha^2 z}\right|_{z=\alpha^{-m}}$$

Evaluating at $m = 3$ and $m = 5$ gives

$$e_3 = \frac{\alpha^4 \alpha^{-3} + \alpha^0}{\alpha^2 \alpha^{-3}} = \frac{\alpha^3}{\alpha^{-1}} = \alpha^4$$
$$e_5 = \frac{\alpha^4 \alpha^{-5} + \alpha^0}{\alpha^2 \alpha^{-5}} = \frac{\alpha^2}{\alpha^{-3}} = \alpha^5$$

The received symbol α^3 at position 3 is therefore corrected to α^6. The received symbol α^4 at position 5 is corrected to α^0. This successfully completes the decoding.

7.6 FREQUENCY DOMAIN ENCODED REED SOLOMON CODES

As the Fourier transform of a Reed Solomon code word contains $n - k$ consecutive zeros, it is possible to encode by considering the information to be a frequency domain vector, appending the appropriate zeros and inverse transforming. The encoding and decoding processes are illustrated in Figure 7.3. In this case the final step of decoding will differ from that in the previous section because we need to obtain the errors in the frequency domain.

We will choose again a (7, 3) double-error correcting RS code over GF(2^3). Let the information be α^2, α^5, α^0 (representing binary information sequence 100111001) and let the zeros in the frequency domain occupy the positions 0 to 3. We therefore put the information into positions 4, 5 and 6 in the frequency domain with zeros in the other positions, producing the transform of the codeword as $\alpha^2 \ \alpha^5 \ \alpha^0 \ 0 \ 0 \ 0 \ 0$.

An inverse Fourier transform generates the codeword as follows:

$$c_0 = \alpha^2 + \alpha^5 + \alpha^0 = \alpha^1$$
$$c_1 = \alpha^2\alpha^1 + \alpha^5\alpha^2 + \alpha^0\alpha^3 = \alpha^0$$
$$c_2 = \alpha^2\alpha^2 + \alpha^5\alpha^4 + \alpha^0\alpha^6 = \alpha^5$$
$$c_3 = \alpha^2\alpha^3 + \alpha^5\alpha^6 + \alpha^0\alpha^2 = \alpha^6$$

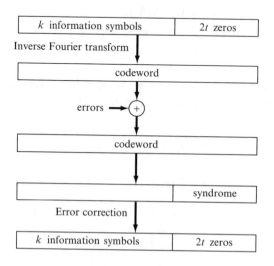

Figure 7.3   Frequency domain encoding and decoding of Reed Solomon codes

$$c_4 = \alpha^2\alpha^4 + \alpha^5\alpha^1 + \alpha^0\alpha^5 = \alpha^5$$
$$c_5 = \alpha^2\alpha^5 + \alpha^5\alpha^3 + \alpha^0\alpha^1 = \alpha^0$$
$$c_6 = \alpha^2\alpha^6 + \alpha^5\alpha^5 + \alpha^0\alpha^4 = \alpha^5$$

Hence the code sequence is α^5 α^0 α^5 α^6 α^5 α^0 α^1. This is in fact the same codeword that we created from different information in Section 7.3. The previous encoding method was systematic whereas this method is not. There is no straightforward way to recover the information from the codeword.

We now create a two-symbol error, say α^5 in position 5 and α^4 in position 3, as in our previous example. The received sequence is α^5 α^4 α^5 α^3 α^5 α^0 α^1.

The decoding proceeds by finding the Fourier transform of the received sequence:

$$R_0 = \alpha^5 + \alpha^4 + \alpha^5 + \alpha^3 + \alpha^5 + \alpha^0 + \alpha^1 = \alpha^0$$
$$R_1 = \alpha^5\alpha^6 + \alpha^4\alpha^5 + \alpha^5\alpha^4 + \alpha^3\alpha^3 + \alpha^5\alpha^2 + \alpha^0\alpha^1 + \alpha^1\alpha^0 = \alpha^1$$
$$R_2 = \alpha^5\alpha^5 + \alpha^4\alpha^3 + \alpha^5\alpha^1 + \alpha^3\alpha^6 + \alpha^5\alpha^4 + \alpha^0\alpha^2 + \alpha^1\alpha^0 = \alpha^0$$
$$R_3 = \alpha^5\alpha^4 + \alpha^4\alpha^1 + \alpha^5\alpha^5 + \alpha^3\alpha^2 + \alpha^5\alpha^6 + \alpha^0\alpha^3 + \alpha^1\alpha^0 = 0$$
$$R_4 = \alpha^5\alpha^3 + \alpha^4\alpha^6 + \alpha^5\alpha^2 + \alpha^3\alpha^5 + \alpha^5\alpha^1 + \alpha^0\alpha^4 + \alpha^1\alpha^0 = \alpha^3$$
$$R_5 = \alpha^5\alpha^2 + \alpha^4\alpha^4 + \alpha^5\alpha^6 + \alpha^3\alpha^1 + \alpha^5\alpha^3 + \alpha^0\alpha^5 + \alpha^1\alpha^0 = \alpha^2$$
$$R_6 = \alpha^5\alpha^1 + \alpha^4\alpha^2 + \alpha^5\alpha^3 + \alpha^3\alpha^4 + \alpha^5\alpha^5 + \alpha^0\alpha^6 + \alpha^1\alpha^0 = \alpha^5$$

Hence the transformed received sequence is α^5 α^2 α^3 0 α^0 α^1 α^0 with the syndrome being 0 α^0 α^1 α^0.

We now form the key equation with $t = 2$ and $\Lambda_0 = 1$:

$$\alpha^0\Lambda_2 + \alpha^1\Lambda_1 + \alpha^0 = 0$$
$$\alpha^1\Lambda_2 + \alpha^0\Lambda_1 + 0 = 0$$

As before the solution is

$$\alpha^6\Lambda = \alpha^1$$
$$\Lambda_1 = \alpha^2 \quad \Lambda_2 = \alpha$$

Instead of carrying out a Chien search, however, we use the connection polynomial to synthesize shift registers with feedback, as in Figure 6.1, to generate the entire frequency domain error sequence. If we initialize the registers with E_2 and E_3 (which are just the syndrome components in those positions), subsequent shifts will generate E_4, E_5 and E_6. This process is known as recursive extension of the frequency domain error sequence. The general arrangement is shown in Figure 7.4 where it is considered that we may have used a method such as the inversionless Berlekamp–Massey algorithm or the Euclidean algorithm to obtain a connection polynomial where $\Lambda_0 \neq 1$.

In our case we have obtained a solution with $\Lambda_0 = 1$; therefore, the circuit becomes as shown in Figure 7.5.

We load the value α^0 into the leftmost stage of the shift register, and zero into the other stage. Cycling the shift register will generate values α^1, α^3, α^3, which are the

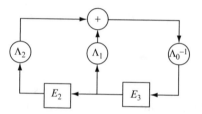

Figure 7.4 General shift register for double-error recursive extension

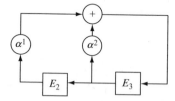

Figure 7.5 Example shift register for recursive extension

frequency domain errors in positions 4, 5 and 6, respectively. The next two values generated are α^0, α^1, corresponding to syndrome components S_0 and S_1. Thus the cycle will repeat; it can be shown that a failure to repeat is the condition for detection of uncorrectable errors.

The complete error sequence in the frequency domain is $\alpha^3 \; \alpha^3 \; \alpha^1 \; 0 \; \alpha^0 \; \alpha^1 \; \alpha^0$. Adding this to $\alpha^5 \; \alpha^2 \; \alpha^3 \; 0 \; \alpha^0 \; \alpha^1 \; \alpha^0$, the Fourier transform of the received signal, gives a decoded sequence of $\alpha^2 \; \alpha^5 \; \alpha^0 \; 0 \; 0 \; 0 \; 0$. Thus we have shown that double-error-correction has been achieved.

7.7 FURTHER EXAMPLES OF REED SOLOMON DECODING

The previous section worked through an example in which the number of errors was exactly equal to the error correcting ability of the code. There are two other types of cases where the outcome is not obvious and which we need to study. They are the cases where the number of errors is less than the error correcting capability and where there are more errors than the code can correct.

Let us first suppose that three errors are introduced into the transmitted codeword and that the received sequence is $\alpha^6 \; \alpha^4 \; \alpha^5 \; \alpha^3 \; \alpha^5 \; \alpha^0 \; \alpha^1$, an extra error of magnitude α having been created in position 6. The Fourier transform is

$$R(z) = \alpha^3 z^6 + \alpha^5 z^5 + \alpha^6 z^4 + \alpha^3 z^3 + \alpha^2 z^2 + \alpha^3 z + \alpha^3$$

from which the key equations are found to be

$$\alpha^2 + \alpha^3\Lambda_1 + \alpha^3\Lambda_2 = 0$$
$$\alpha^3 + \alpha^2\Lambda_1 + \alpha^3\Lambda_2 = 0$$

Eliminating Λ_2 gives

$$\alpha^5 + \alpha^5\Lambda_1 = 0$$
$$\Lambda_1 = \alpha^0$$
$$\Lambda_2 = \alpha^2$$

The registers for recursive extension are shown in Figure 7.6. After initializing with values α^3 α^2, the error sequence generated is α^2 α^6 α^5 followed by α^5. The sequence is therefore incorrect because the syndrome s_0 has not been generated at this point. The general condition for detection of uncorrectable errors is that the sequence generated by recursive extension is not cyclic with the correct length. This can be shown to be identical to the condition that the Chien search does not return the correct number of roots.

Looking next at the case where we have fewer than the maximum number of correctable errors, it is obvious that if there are no errors then the transform of the codeword will exhibit zeros in the expected places and no decoding is required. If there are some errors, but less than t, the transform of the error locator polynomial will have fewer than t unknown roots and the t simultaneous equations will not be linearly independent. This will be illustrated by returning to the example of the previous section but this time introducing only one error.

Let us assume that we receive a sequence α^5 α^4 α^5 α^6 α^5 α^0 α^1. This is the same as the previous example except that position 3 does not contain an error, the sole error being in position 5. The transform is

$$R(z) = \alpha^6 z^6 + \alpha^3 z^5 + \alpha^5 z^4 + \alpha^6 z^3 + \alpha^1 z^2 + \alpha^3 z + \alpha^5$$

from which the key equations are found to be

$$\alpha^1 + \alpha^3\Lambda_1 + \alpha^5\Lambda_2 = 0$$
$$\alpha^6 + \alpha^1\Lambda_1 + \alpha^3\Lambda_2 = 0$$

This is similar to what happened in the single-error example of Section 6.9. The second equation is the same as the first, multiplied by α^2, which tells us that there is

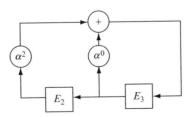

Figure 7.6 Attempted recursive extension after triple error

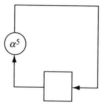

Figure 7.7   Recursive extension with single error

only one error and we must set Λ_2 equal to zero. The solution to the key equation is
thus

$$\Lambda_1 = \alpha^5$$

and loading the value α^6 into the shift register of Figure 7.7 gives the sequence
$\alpha^4 \ \alpha^2 \ \alpha^0$ as the corrections to the frequency domain data positions 4, 5 and 6, then
the next value is α^5 which reproduces the value of s_0, showing that the error correction is successful. We therefore correct the received Fourier transform values in
positions 6, 5 and 4, $\alpha^6 \ \alpha^3 \ \alpha^5$ to $\alpha^2 \ \alpha^5 \ \alpha^0$, the information encoded in Section 7.7.
   The fact that in the above example each term in the syndrome is a constant factor
times the previous term is characteristic of single symbol errors, and the position of
the error in the time domain can be readily determined from the factor. In this case
because $S_j = \alpha^5 S_{j-1}$, the error is in position 5 and the value of s_0 is the error value in
that location. It is often worth including a check for single errors in a decoder and a
special decoding routine, because in many practical examples single errors will make
up a significant proportion of the sequence errors. Moreover all decoders will have to
cope with the case where there are fewer than t errors.

7.8   ERASURE DECODING

One feature that is sometimes incorporated into decoders for Reed Solomon codes is
the ability to recover events known as erasures. These are instances when there is
knowledge that a symbol is likely to be in error, perhaps through the detection of
interference. Erasure decoding could be regarded as a first step in soft-decision
decoding because, in comparison with what was transmitted, an extra level has
been introduced into the received sequence.
   When an erasure occurs, the maximum likelihood decoding method is to compare
the received sequence with all codewords, but ignoring the symbol values in the
erased positions. The erasures are then filled using the values from the selected
codeword. With e erasures there will still be a minimum distance of $d_{min} - e$ between
codewords, counting only the unerased places. Thus we will obtain decoding provided $2t$ is less than this reduced minimum distance:

$$2t + e < d_{min} \tag{7.3}$$

At this point, one might well ask whether there is any point in declaring erasures on a binary symmetric channel. A t-error correcting code will be able to fill up to $2t$ erasures, but if a guess was made for the values of all the erased symbols then on average half would be correct and the error correction would cope with the rest. It will be seen shortly that erasures do give some advantage, in that erasure filling is equivalent to carrying out error correction on two possible sets of reconstituted bit values, and then choosing the better of the two. On the other hand, the demodulator need only have a slight inclination towards one or other value to make choosing the more likely bit value a better strategy than erasure.

For any binary code there is a straightforward, nonalgebraic erasure filling method. Replace all the erased bits by zero and decode. If no more than half of the erasures should have been ones and Equation (7.3) was satisfied, then the number of errors will still be less than half of d_{min} and the decoding will be correct. If on the other hand more than half the erasures should have been ones then we may get a decoding that will introduce extra errors into the sequence.

In this case, replacing all the erased bits by 1 will be successful. The procedure is therefore to decode twice, replacing all the erased bits firstly with zeros and then with ones. If the decoded sequences differ, choose the one that is closer to the received sequence.

In contrast, if several bits of a Reed Solomon symbol are erased, it is unlikely that they can be guessed correctly and erasure filling is a good strategy. They may be decoded in the presence of erasures by an algebraic technique to be explained below. The minimum distance of these codes is $n - k + 1$, which means that in the absence of errors, Equation (7.3) shows that $n - k$ erasures can be filled. We thus have the interesting result that a Reed Solomon codeword can be recovered from any k correct symbols.

To decode algebraically, we replace the erased symbols by some arbitrary value, usually zero. We adopt the polynomial approach in the frequency domain and multiply the product of the syndrome polynomial $S(z)$ and the connection polynomial $\Lambda(z)$ by an erasure polynomial $\Gamma(z)$ which is known because the positions of the erasures are known. For every two erasures, the degree of the error locator polynomial is reduced by one so that the degree of the product of erasure and error locator polynomials is increased by one. The number of simultaneous equations that can be formed will thus be reduced by one, thus matching the reduction in the number of unknowns. This all sounds rather horrific, but is fairly straightforward if considered in the context of an example.

7.9   EXAMPLE OF ERASURE DECODING OF REED SOLOMON CODES

We choose as our example the first case from Section 7.7 in which there was one error at position 5 in the received sequence, but we introduce also two erasures in positions 6 and 1. The received sequence will be taken as $0\ \alpha^4\ \alpha^5\ \alpha^6\ \alpha^5\ 0\ \alpha^1$. The Fourier transform is

$$R(z) = \alpha^6 z^6 + \alpha^6 z^5 + \alpha^3 z^4 + \alpha^1 z^3 + \alpha^6 z^2 + \alpha^5 + \alpha^0$$

The low order terms of $R(z)$ form a syndrome of the received sequence. The erasure polynomial is

$$\Gamma(z) = (\alpha^6 z + 1)(\alpha^1 z + 1) = z^2 + \alpha^5 z + 1$$

The error locator polynomial is

$$\Lambda(z) = \Lambda_1 z + 1$$

and the product is

$$\Gamma(z)\Lambda(z) = \Lambda_1 z^3 + [\alpha^0 + \alpha^5 \Lambda_1] z^2 + [\alpha^5 + \Lambda_1] z + 1$$

As this polynomial is of degree 3, we can only carry out a single place convolution with a known section of the error spectrum to produce a key equation. We are therefore only interested in the terms of degree 3 when we multiply by the syndrome

$$S(z) = \alpha^1 z^3 + \alpha^6 z^2 + \alpha^5 z + \alpha^0$$

giving as our key equation

$$\alpha^0 \Lambda_1 + \alpha^5 + \alpha^3 \Lambda_1 + \alpha^4 + \alpha^6 \Lambda_1 + \alpha^1 = 0$$

$$\alpha^5 \Lambda_1 = \alpha^3; \quad \Lambda_1 = \alpha^5$$

We can now substitute this value back into the expression for $\Gamma(z)\Lambda(z)$, which will be the polynomial used to generate the errors in the frequency domain by recursive extension:

$$\Gamma(z)\Lambda(z) = \alpha^5 z^3 + \alpha^1 z^2 + 1$$

This gives rise to the circuit shown in Figure 7.8 to be used for recursive extension. Loading with values α^5 α^6 α^1 and shifting gives the sequence α^1 α^1 α^0 and then, regenerating the syndrome, α^0 α^5, α^6. The error correction is therefore successful and the values α^0 α^1 α^1 are added to the components α^6 α^6 α^3 from the Fourier transform of the received sequence to give recovered information α^2 α^5 α^0.

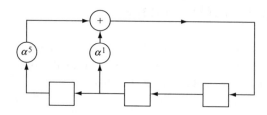

Figure 7.8   Recursive extension with one error and two erasures

If instead we choose to use the Forney algorithm to correct errors in the time domain, the error evaluator polynomial is found to be

$$\Omega(z) = [S(z)\Gamma(z)\Lambda(z)] \bmod z^4 = \alpha^5 z^2 + \alpha^5 z + \alpha^0$$

Note that the degree of this polynomial is now 2 because there are effectively three errors to correct. To calculate error values, we also need the formal derivative of $\Gamma(z)\Lambda(z)$, i.e. of $\alpha^5 z^3 + \alpha^1 z^2 + \alpha^0$. The value of this derivative is $\alpha^5 z^2$. Therefore at position i, the error value is

$$e_i = \left. \frac{\alpha^5 z^2 + \alpha^5 z + \alpha^0}{\alpha^5 z^3} \right|_{z = \alpha^{-i}}$$

The polynomial $\Gamma(z)\Lambda(z)$ has known roots α^{-6} and α^{-1} (because erasures were detected there) and a further root α^{-5} that can be located by Chien search. Calculating the error values at positions 1, 5 and 6 gives $e_1 = \alpha^0$, $e_5 = \alpha^5$, $e_6 = \alpha^5$. The received sequence is therefore corrected to $\alpha^5 \ \alpha^0 \ \alpha^5 \ \alpha^6 \ \alpha^5 \ \alpha^0 \ \alpha^1$, corresponding to the original codeword.

Euclid's algorithm may be used to solve for the feedback polynomial in the presence of erasures by initializing $g_1(z)$ as $\Gamma(z)$ and terminating when the degree of the remainder is less than $t + e/2$ (e is the degree of the erasure locator polynomial, assumed to be even). To operate the Berlekamp–Massey algorithm, set l and n to e and initialize $\Lambda(z)$ to $\Gamma(z)$.

7.10   GENERALIZED MINIMUM DISTANCE DECODING

Having the ability to fill in erasures allows us to deal effectively with certain types of interference and also to implement some approaches to soft-decision decoding. One well-known approach is Forney's *Generalized Minimum Distance* (GMD) decoding [1].

The principle of GMD decoding is very simple. We use a t-error correcting code and, from the soft-decision information on the received sequence, we rank the received symbols according to reliability. The decoding procedure is then:

1   Decode for t-error correction.

2   Erase the two least reliable symbols and decode for $t - 1$ errors and 2 erasures.

3   If the number of erasures is $2t$, stop decoding and go to step 5.

4   From the symbols input to the previous decoding, erase the two least reliable (i.e. increase the number of erasures by 2), decode and return to step 3.

5   From the decoded sequences produced, choose the closest to the received sequence.

There will be at most $t + 1$ decoded sequences to compare; in practice when the number of errors exceeds t there may be fewer because the errors are likely to be uncorrectable. It is not clear, however, whether the performance gains will be worth the effort of carrying out $t + 1$ decoding attempts.

The GMD algorithm is designed to work in conjunction with a particular metric of goodness of decoded solutions. For each received symbol i with reliability r_i, we count $+r_i$ if the codeword matches and $-r_i$ if it does not. There can be at most one codeword for which the sum over all symbols will exceed $n - d_{min}$ and, if there is such a codeword, GMD decoding will find it. For a binary code this is a good metric, although GMD is still a bounded distance decoding rather than true maximum likelihood. For multilevel codes transmitted over a binary channel, however, the metric may not be good. Over a memoryless channel, we would consider that two symbol values differing in a single bit are closer than two symbol values differing in several bits. The GMD metric, however, treats those two cases as the same because symbols are treated as being either the same or different. Only in burst-error conditions, or where an orthogonal modulation such as MFSK is used, will the GMD metric be reasonable for a Reed Solomon code.

If we do, however, have conditions in which GMD seems to offer performance benefits, the complexity needs to be considered. Using the Berlekamp–Massey or the Euclidean algorithm, every decoding is a separate exercise with no way of reusing results from previous decodings. However GMD can be implemented in conjunction with another algorithm, the *Welch–Berlekamp algorithm*, in a way that builds on previous results.

7.11   WELCH–BERLEKAMP ALGORITHM

The Welch–Berlekamp algorithm is another algebraic method for decoding Reed Solomon codes. It is a time domain algorithm using values that are easily computed from the received sequence, rather than the frequency domain syndromes used by other algebraic methods. As a result, although the Welch–Berlekamp algorithm itself is slightly more complex than the Berlekamp–Massey algorithm, the overall complexity may well be lower if the actual number of errors to be corrected is not too large. Moreover, it fits conveniently with GMD decoding.

The Welch–Berlekamp algorithm can be thought of as a kind of curve fitting process. The points each have a value and a position, and a curve can be constructed to fit any k points. When two more points are added, the curve must fit at least $k + 1$, but the curve is allowed to miss one of the points. After adding another two points, the curve must fit at least $k + 2$ of them. When eventually all n points have been considered, the curve must fit at least $(n + k)/2$ of them.

Suppose, as in our previous double-error correction examples we transmit a codeword from a (7, 3) RS code with roots α^3, α^2, α^1, α^0 in GF(8). The first step is to precompute some values needed for input to the algorithm and for error value calculations. For the input to the algorithm we evaluate

$$\sum_{j=0}^{n-k-1} g_j x^j = \frac{g(X)}{X + \alpha^{n-k-1}} = (X + \alpha^0)(X + \alpha^1)(X + \alpha^2) = X^3 + \alpha^5 X^2 + \alpha^6 X + \alpha^3$$

Therefore $g_3 = \alpha^0$, $g_2 = \alpha^5$, $g_1 = \alpha^6$, $g_0 = \alpha^3$.

For error value calculations we compute $C = \alpha^0\alpha^1 \cdots \alpha^{n-k-2}(\alpha^0 + \alpha^1)(\alpha^0 + \alpha^2) \cdots (\alpha^0 + \alpha^{n-k-1})$. For our example code the answer is α^6. We now need for each of the data locations the value of $h_i = C/g(\alpha^i)$ where $g(\alpha^i)$ indicates the evaluation of $g(X)$ at each of the data locations. The values are found to be

$$h_6 = \alpha^6/\alpha^4 = \alpha^2$$
$$h_5 = \alpha^6/\alpha^1 = \alpha^5$$
$$h_4 = \alpha^6/\alpha^0 = \alpha^6$$

Assume we receive a sequence $\alpha^5 \; \alpha^4 \; \alpha^5 \; \alpha^3 \; \alpha^5 \; \alpha^0 \; \alpha^1$. The first step is to compute the syndrome obtained by division by the generator. This is found to be $s_3 = \alpha^0$, $s_2 = \alpha^2$, $s_1 = \alpha^0$, $s_0 = \alpha^6$.

The input to the Welch–Berlekamp algorithm is the set of points (S_j, α^j) where $S_j = s_j/g_j$. The input points are therefore (α^0, α^3), (α^4, α^2), (α^1, α^1), (α^3, α^0).

We need to find two polynomials $Q(x)$ and $N(x)$ for which

$$Q(\alpha^j)S_j = N(\alpha^j) \quad \text{for } 0 \leq j \leq n - k - 1$$

and the length $L[Q(x), N(x)]$, defined as the maximum of $\deg[Q(x)]$ and $\deg[N(x)] + 1$, has the minimum possible value.

The steps in the algorithm are now:

1 Set $Q^0(x) = 1$, $N^0(x) = 0$, $W^0(x) = x$, $V^0(x) = 1$ and $d = 0$.

2 Evaluate $D_1 = Q^d(\alpha^d)S_d + N^d(\alpha^d)$.

3 If $D_1 = 0$, set $W^{d+1} = W^d(x + \alpha^d)$, $V^{d+1} = V^d(x + \alpha^d)$ and go to step 6; otherwise, set $D_2 = W^d(\alpha^d)S_d + V^d(\alpha^d)$.

4 Set $Q^{d+1} = Q^d(x + \alpha^d)$, $N^{d+1} = N^d(x + \alpha^d)$, $W^{d+1} = W^d + Q^d D_2/D_1$, $V^{d+1} = V^d + N^d D_2/D_1$.

5 Check whether $L[W^d, V^d]$ was less than or equal to $L[Q^d, N^d]$; if it was then swap Q^{d+1}, N^{d+1} with W^{d+1}, V^{d+1}.

6 Increment d.

7 If $d < n-k$, return to step 1; otherwise, $Q(x) = Q^d(x)$, $N(x) = N^d(x)$.

The steps in evaluation for this example are shown in Table 7.3.

The error locator polynomial is $x^2 + \alpha^2 x + \alpha^1$ which has roots α^3 and α^5. Finding two roots for a degree 2 polynomial indicates that errors are correctable.

The error values in the data positions are

$$e_k = h_k \frac{N(\alpha^k)}{Q'(\alpha^k)} \tag{7.4}$$

Table 7.3

d	Input	D_1	D_2	$L_1 \leq L_2$	$Q(x)$	$N(x)$	$W(x)$	$V(x)$
					1	0	x	1
0	α^3	α^3	α^1	N	$x + \alpha^0$	0	$x + \alpha^5$	1
1	α^1	α^4	0	Y	$x + \alpha^5$	1	$x^2 + \alpha^3 x + \alpha^1$	0
2	α^4	0					$x^3 + \alpha^5 x^2 + \alpha^6 x + \alpha^3$	0
3	α^4	α^4	α^2	N	$x^2 + \alpha^2 x + \alpha^1$	$x + \alpha^3$	$x^3 + \alpha^5 x^2 + \alpha^1 x$	α^5

where $Q'(x)$ is the formal derivative of $Q(x)$ (see Section 7.4). In this case $Q'(x) = \alpha^2$. Therefore the error value at location 5 is $e_5 = \alpha^5(\alpha^5 + \alpha^3)/\alpha^2 = \alpha^5$. The received information is therefore corrected to $\alpha^5\ \alpha^0\ \alpha^5$.

The application of the Welch–Berlekamp algorithm to GMD is that, for a Reed Solomon code, any $n - k$ symbols can be treated as parity checks. The least reliable $n - k$ symbols are therefore treated as parity checks with the syndromes derived by recomputing the codeword from the most reliable symbols. As we enter the syndromes into the algorithm, each time a new pair of values of Q and N is produced we carry out the correction and produce a new codeword for consideration.

7.12   SINGLY EXTENDED REED SOLOMON CODES

It is possible to create a q-ary Reed Solomon code of length q, and such a code is known as an extended code. The code will still correct $t = (n - k)/2$ symbols and can be thought of either as adding a parity symbol to a code which corrects $t - 1$ errors and detects t (expansion), or as adding an extra information symbol to a t-error correcting code. The extended code is not cyclic, but can be encoded and decoded using frequency domain techniques. The properties of such a code are relatively easy to understand and the logic of the decoding approach is effectively the proof of the properties.

To create a t-error correcting RS code of length q, we first create the code of length $q - 1$ with $2t - 1$ parity checks. The frequency domain syndrome components s_0 to s_{2t-2} will, by definition, be zero. We now choose an adjacent position in the frequency domain, known as the edge frequency, work out the value of Fourier transform in that position and append it to the sequence. This is shown in Figure 7.9 on the assumption that the edge frequency corresponds to s_{2t-1}. The extra symbol is known as the edge symbol.

In the frequency domain, a codeword can be created as shown in Figure 7.10. Here one of the information symbols acts as the edge symbol.

To decode, we calculate $2t$ components of the frequency domain syndrome of the length $q - 1$ sequence. We use $2t - 2$ syndrome components to create a connection polynomial of degree up to $t - 1$. If there are no more than $t - 1$ errors in the length $q - 1$ sequence then the final syndrome component s_{2t-2} will show that there are no

s_{2t-1}	k information symbols	$2t-1$ parity symbols

Figure 7.9   Singly extended Reed Solomon code

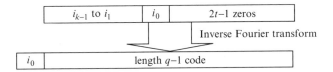

Figure 7.10 Frequency domain view of singly extended Reed Solomon code

more errors detected and correction can proceed. If there are t errors in the length $q - 1$ sequence then they will be detected by s_{2t-2}; however, in this case we can assume that the received value of s_{2t-1} is correct (otherwise there would be more than t errors in the whole codeword). We therefore add the received value of s_{2t-1} to the calculated value to produce a usable value for t-error correction.

The choice of edge frequency from the two available is unimportant for the properties of the code. However, assuming that the Berlekamp–Massey algorithm is used for decoding, the choice of s_{t-2} will be convenient as the connection polynomial can, if necessary, be calculated as an extension of the results up to s_{2t-2}. The location of the edge symbol in the final codeword is entirely arbitrary.

Examples of singly extended code

To create a single-error correcting double-error detecting RS code of length 7 over GF(8), we can use

$$g(X) = (X + \alpha^0)(X + \alpha^1)(X + \alpha^2) = X^3 + \alpha^5 X^2 + \alpha^6 X + \alpha^3$$

The information $\alpha^5 \ \alpha^0 \ \alpha^5 \ \alpha^2$ encodes systematically to $\alpha^5 \ \alpha^0 \ \alpha^5 \ \alpha^2 \ 0 \ \alpha^2 \ \alpha^0$. The Fourier transform in position 3 is

$$s_3 = \alpha^5 \alpha^4 + \alpha^0 \alpha^1 + \alpha^5 \alpha^5 + \alpha^2 \alpha^2 + \alpha^2 \alpha^3 + \alpha^0 = \alpha^6$$

The codeword is therefore $\alpha^6 \ \alpha^5 \ \alpha^0 \ \alpha^5 \ \alpha^2 \ 0 \ \alpha^2 \ \alpha^0$.

Now create errors in positions 7 and 5, the received sequence being $\alpha^2 \ \alpha^5 \ \alpha^4$ $\alpha^5 \ \alpha^2 \ 0 \ \alpha^2 \ \alpha^0$. The syndrome polynomial of the sequence $\alpha^5 \ \alpha^4 \ \alpha^5 \ \alpha^2 \ 0 \ \alpha^2 \ \alpha^0$ is

$$s(z) = \alpha^1 z^2 + \alpha^3 z + \alpha^5$$

and the syndrome in position 4 is zero. Attempting single-error correction yields a connection polynomial of $\alpha^5 z + 1$ and the syndrome component in position 2 is correctly predicted. The symbol in position 7 is therefore not needed and can be discarded. The error value is the value of s_0, i.e. α^5, so the received word of the (7, 4) code can be corrected to $\alpha^5 \ \alpha^0 \ \alpha^5 \ \alpha^2 \ 0 \ \alpha^2 \ \alpha^0$, with the first four symbols being the information.

Alternatively, create errors in positions 5 and 3, the received sequence being $\alpha^1 \ \alpha^5 \ \alpha^4 \ \alpha^5 \ \alpha^1 \ 0 \ \alpha^2 \ \alpha^0$. The syndrome of $\alpha^5 \ \alpha^4 \ \alpha^5 \ \alpha^1 \ 0 \ \alpha^2 \ \alpha^0$ is

$$s(z) = \alpha^0 z^2 + \alpha^1 z + \alpha^0$$

with the value in position 3 being α^6. The single-error correcting connection polynomial is $\alpha^1 z + 1$, but this does not correctly predict the next component. We therefore assume that the edge frequency α^6 was correct and add it to the calculated s_3 value to give a syndrome for a double-error correcting code

$$s(z) = \alpha^0 z^2 + \alpha^1 z + \alpha^0$$

The key equation for double error correction is

$$\alpha^0 + \alpha^1 \Lambda_1 + \alpha^0 \Lambda_2 = 0$$

$$\alpha^0 \Lambda_1 + \alpha^1 \Lambda_2 = 0$$

Eliminating Λ_2 gives $\alpha^1 + \alpha^6 \Lambda_1 = 0$, $\Lambda_1 = \alpha^2$, $\Lambda_2 = \alpha^1$. As in Section 7.5, the roots are α^2 and α^4 indicating errors in positions 5 and 3. The error values may be found using the Forney algorithm.

7.13   DOUBLY EXTENDED REED SOLOMON CODES

The single extension of RS codes can be taken one stage further to create a code of length $q + 1$. A $t - 1$-error correcting code is created, two edge frequencies selected and the Fourier transform calculated for those two frequencies to create the two edge symbols. The decoding process is slightly more complicated than for the singly extended code. The logic, assuming that no more than t errors have occurred is as follows:

If both the edge symbols are incorrect then no more than $t - 2$ errors need be corrected in the length $q - 1$ codeword. This can be achieved from the normal parity symbols with two more parities available to check the correctness of the process. Therefore if correction of up to $t - 2$ errors appears to be successful (as verified by the remaining parity checks), accept the result.

If one edge symbol is incorrect then no more than $t - 1$ errors need be corrected in the length $q - 1$ codeword. This can be achieved from the normal parity symbols. Therefore if correction of up to $t - 1$ errors appears to be successful, verify this by calculating the two edge symbols and checking that at least one of them corresponds to what was received. If this condition is satisfied then accept the result.

For t-error correction to be needed in the length $q - 1$ codeword, both edge symbols must be correct. Therefore if decoding has so far been unsuccessful, assume that both edge symbols are correct and add them to the received sequence Fourier transform values at the corresponding edge frequencies. Use the results as two more syndrome components to achieve t-error correction.

Again the edge frequencies could be chosen at either end of the syndrome frequencies, or even one at each end. However, if the syndromes are s_0 to s_{2t-3}, choosing s_{2t-2} and s_{2t-1} as the edge symbols will again fit conveniently with the operation of the Berlekamp–Massey algorithm.

To make the point that extended codes can be viewed as adding information symbols to a t-error correcting code and that other positions can be chosen for the edge frequencies, consider a doubly extended RS code whose frequency domain view is shown in Figure 7.11. The process of encoding is as follows. Put $k - 2$ ($= q - 1 - 2t$) information symbols and $2t$ consecutive zeros into a vector of length $q - 1$. Put the two remaining information symbols into the two outside zero positions, one at the low-order end and one at the high-order end. We now have the spectrum of a $t - 1$ error-correcting Reed Solomon code. The vector of length $q - 1$ is given an inverse Fourier transform over $GF(q)$ to produce a Reed Solomon codeword which can correct $t - 1$ errors. The high-order edge frequency is appended at the beginning (high-order end) of the codeword and the low-order edge frequency at the (low-order) end of the codeword. The codeword is now of length $q + 1$.

To decode a received sequence using the extended code, strip the symbols from the beginning and end and forward transform the remaining sequence of length $q - 1$. The syndrome consists of the high-order edge frequency plus the stripped symbol received in the high-order position, the subsequent $2t - 2$ symbol values and the sum of the low-order edge frequency plus the stripped symbol received in the low-order position. This is shown in Figure 7.12. Decoding, however, starts off with an attempt at $t - 1$ error correction, using only the central $2t - 2$ symbols of the syndrome following the logic outlined previously.

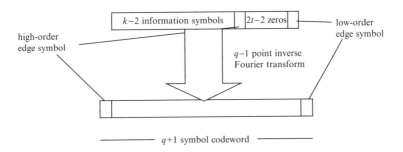

Figure 7.11   Encoding of doubly extended Reed Solomon code

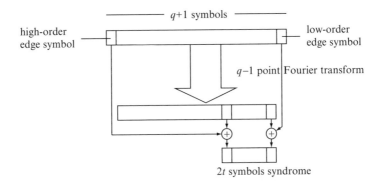

Figure 7.12   Syndrome formation for doubly extended Reed Solomon code

Example of doubly extended code

We consider the case of a (9, 5) Reed Solomon code over GF(8). In this case we have a 5 symbol information sequence 1 α^3 α α^6 α^2. We treat the first and last of these as edge symbols, placing them in positions 3 and 0, respectively, in the frequency domain. Thus we start from the spectrum of a single-error correcting Reed Solomon code

$$C(z) = \alpha^3 z^6 + \alpha z^5 + \alpha^6 z^4 + z^3 + \alpha^2$$

The inverse transform gives the codeword of a (7, 5) RS code:

$$c(X) = \alpha^6 X^6 + \alpha^2 X^5 + \alpha^3 X^4 + \alpha^6 X^3 + X^2 + \alpha^3 X + 1$$

The two extra information symbols are now added to the ends of the codeword to produce the (9, 5) RS codeword

$$c'(X) = X^8 + \alpha^6 X^7 + \alpha^2 X^6 + \alpha^3 X^5 + \alpha^6 X^4 + X^3 + \alpha^3 X^2 + X + \alpha^2$$

Now assume that errors occur at positions 8 and 5 producing a received sequence

$$r'(X) = \alpha^3 X^8 + \alpha^6 X^7 + \alpha^2 X^6 + \alpha^5 X^5 + \alpha^6 X^4 + X^3 + \alpha^3 X^2 + X + \alpha^2$$

The additional symbols are stripped from the sequence to give

$$c(X) = \alpha^6 X^6 + \alpha^2 X^5 + \alpha^5 X^4 + \alpha^6 X^3 + X^2 + \alpha^3 X + 1$$

which transforms to

$$R(z) = \alpha^2 z^6 + \alpha^3 z^4 + \alpha^3 z^2 + \alpha^6 z$$

The terms of order 3 downwards are extracted as the syndrome, with the additional symbols added at the appropriate points, giving

$$R(z) = z^3 + \alpha^3 z^2 + \alpha^6 z + \alpha^2$$

The terms in the syndrome are incrementing by multiples of α^4 as the order increases, indicating the equivalence of a single error with the key equation solving as $\Lambda^1 = \alpha^4$. The spectrum of the error sequence is therefore

$$E(z) = \alpha^5 z^6 + \alpha z^5 + \alpha^4 z^4 + z^3 + \alpha^3 z^2 + \alpha^6 z + \alpha^2$$

which, when added to $R(z)$ gives

$$C(z) = \alpha^3 z^6 + \alpha z^5 + \alpha^6 z^4 + z^3 + \alpha^2$$

Decoding has been correctly achieved and all the information is directly available from $C(z)$.

7.14 CONCLUSION

Reed Solomon codes are arguably the most elegant and most useful product of algebraic coding theory. The origin and the discovery of their relation to BCH codes were mentioned in Chapter 6. The implementations of algorithms mentioned in that chapter, and of finite field arithmetic from Chapter 5 are also highly relevant. The Welch–Berlekamp algorithm was originally embodied in a US patent [2]. The anticipated expiry of that patent gave rise to considerable interest and the publication of several papers describing it and its uses, of which [3] describes its application to GMD decoding. A similar method described in [4] includes a method to keep track of the results of the Chien search to assist the multiple decoding attempts. Some applications of Reed Solomon codes will be discussed in Chapters 9 and 10.

7.15 EXERCISES

1   Find the generator polynomial of the double-error-correcting RS code of length 7 whose Fourier transform has zeros in positions 0, 1, 2 and 3, assuming that GF(8) is generated by $X^3 + X^2 + 1$. Find the syndrome of the sequence $\alpha^2 \ \alpha^4 \ 0 \ \alpha^6$ $\alpha^6 \ \alpha^5 \ \alpha^6$ by long division and by computation of the appropriate Fourier transform components.

2   Using the (7, 3) RS code example of question 1, carry out the decoding if the errors are

$$e(X) = \alpha^5 X^4 + \alpha^2 X^2$$

using Euclid's algorithm for solution of the key equation.

3   Using the RS code of the above question, show the decoding with a single error affecting all bits in position 4.

4   In the example (7, 3) Reed Solomon code of Section 4.16, erase the transmitted symbols at positions 6, 5, 3 and 0. Carry out the decoding.

5   Encode the 8-ary information sequence $\alpha^2 \ 0 \ \alpha^6 \ 1 \ \alpha$ into a (9, 5) extended Reed Solomon code. Carry out the decoding if errors are introduced as follows:

1 in position 8 and α in position 0
α^2 in position 7 and α^3 in position 2

7.16 REFERENCES

1 D. Forney, *Generalized minimum distance decoding*, IEEE Transactions on Information Theory, Vol. IT-12, pp. 125–131, 1966.

2 R. Berlekamp and L.R. Welch, *Error correction for algebraic block codes*, US Patent no. US4633470, December 1986.

3 E. Peile, *On the performance and complexity of a generalized minimum distance Reed-Solomon decoding algorithm*, International Journal of Satellite Communications, Vol. 12, pp. 333–359, 1994.

4 K. Sorger, *A new Reed-Solomon code decoding algorithm based on Newton's interpolation*, IEEE Transactions on Information Theory, Vol. 39, pp. 358–365, March 1993.

8

Performance calculations for block codes

8.1 INTRODUCTION

The consideration of block codes has largely concentrated on the implementation. The purpose of this chapter is to look at the performance of the codes, both for error detection and for error correction.

Common sense tells us that for given values of n and k there must be a limit to the minimum distance that can be obtained, however it is interesting to know what the limitations are. It is found that there is no fixed relationship between the code parameters, but there are several upper bounds applying to minimum distance or error correction (Hamming bound, Plotkin bound, Griesmer bound and Singleton bound) and one lower bound (Gilbert–Varsharmov bound) which tells us a value of minimum distance that we should be able to achieve.

In this chapter we start by studying the above-mentioned bounds and then go on to consider how to carry out performance calculations for particular codes.

8.2 HAMMING BOUND

We have already met in Chapter 3 the Hamming bound, which states that the number of syndromes is at least equal to the number of correctable error patterns. For a q-ary symbol, any error can have $q - 1$ possible values, and the formula becomes:

$$q^{n-k} \geq 1 + n(q - 1) + \frac{n(n - 1)}{2}(q - 1)^2 + \frac{n(n - 1)(n - 2)}{3}(q - 1)^3 + \cdots$$

$$q^{n-k} \geq \sum_{i=0}^{t} \begin{bmatrix} n \\ i \end{bmatrix}(q - i)^i \tag{8.1}$$

The Hamming bound may be used to obtain a useful approximation for the probability of incorrect output when the weight of the errors exceeds the capability of a bounded distance decoder. The decoder will produce incorrect output if and only if the syndrome produced is one that corresponds to a correctable error. If the syndromes of uncorrectable errors are considered to be uniformly distributed over all possible values, the probability of miscorrection is

$$P_{de} = \frac{\sum_{i=0}^{t} \left[\begin{matrix} n \\ i \end{matrix} \right] (q-i)^i}{q^{n-k}} \tag{8.2}$$

Of course the syndromes of uncorrectable errors are not uniformly distributed because usually the weights of the errors are not uniformly distributed and there will be more chance of a weight $t+1$ error than of a higher weight error. An exact evaluation is possible, but extremely complex, provided the weight distribution of the code is known. However, for the codes of interest in this book, Equation (8.2) can be considered to give an upper bound to the probability of incorrect output, given that an uncorrectable error has occurred.

8.3   PLOTKIN BOUND

The Plotkin bound is similar to the Hamming bound, in that it sets an upper limit to d_{min} for fixed values of n and k. It tends however to set a tighter bound for low rate codes, the Hamming bound being tighter for higher rates.

The Plotkin bound applies to linear codes and states that the minimum distance is at most equal to the average weight of all nonzero codewords. For a q-ary code with n symbols, the chance over the whole set of codewords of any symbol being nonzero is $(q-1)/q$ (provided the code is linear) and there are q^k-codewords in the whole set. The number of nonzero codewords is $q^k - 1$ and so the average weight of a codeword is

$$\frac{n \frac{q-1}{2} q^k}{q^k - 1}$$

The minimum distance cannot be greater than this, so

$$d_{min} \le \frac{n(q-1)q^{k-1}}{q^k - 1} \tag{8.3}$$

For a binary code this becomes

$$d_{min} \le \frac{n2^{k-1}}{2^k - 1} \tag{8.4}$$

It is not easy to find the maximum value of k for a given n and d_{min}, but it can be shown from the above result that

$$k \le n \frac{q d_{min} - 1}{q - 1} + 1 + \log_q d_{min} \tag{8.5}$$

or for a binary code

$$k \leq n - 2d_{min} + 2 + \log_2 d_{min} \tag{8.6}$$

8.4 GRIESMER BOUND

The Griesmer bound is often tighter than the Plotkin bound, and its derivation leads to methods of constructing good codes. Let $N(k, d)$ represent the lowest possible value of length n for a linear code C of dimension k and minimum distance d. Without loss of generality, the generator matrix can be taken to have a first row consisting of d ones followed by $N(k, d) - d$ zeros:

$$\mathbf{G} = \begin{bmatrix} 111\ldots1 & 000\ldots0 \\ \mathbf{G_1} & \mathbf{G_2} \end{bmatrix}$$

The matrix $\mathbf{G_2}$ generates a $(N(k, d) - d, k - 1)$ code of minimum distance d_1, called the residual code. If $\mathbf{u}$ is a codeword of the residual code which, when concatenated with a sequence $\mathbf{v}$ of length d, produces a codeword of C, then we can say

$$d_1 + \text{weight}(\mathbf{v}) \geq d$$

However $\mathbf{u}$ concatenated with the complement of $\mathbf{v}$ is also a codeword:

$$d_1 + d - \text{weight}(\mathbf{v}) \geq d$$

Therefore $2d_1 \geq d$ or $d_1 \geq \lceil d/2 \rceil$ (the symbol $\lceil d/2 \rceil$ represents the integer which is not less than $d/2$). Since the code generated by $\mathbf{G_2}$ is of length $N(k, d) - d$, we can say

$$N(k, d) = N(k - 1, \lceil d/2 \rceil) + d$$

Applying this result iteratively gives

$$N(k, d) = \sum_{i=0}^{k-1} \frac{\lceil d \rceil}{2^i}$$

This is the lowest possible value of length, so the general statement of the Griesmer bound for binary codes is

$$n \geq \sum_{i=0}^{k-1} \frac{\lceil d \rceil}{2^i} \tag{8.7}$$

For q-ary codes, the argument generalizes to give

$$n \geq \sum_{i=0}^{k-1} \frac{\lceil d \rceil}{q^i} \tag{8.8}$$

8.5   SINGLETON BOUND

If we change one of the information symbols in a block code, the best we can hope for in terms of distance between codewords is that all the parity symbols will also change. In this case the distance between the two codewords will be $n - k + 1$. This sets an upper bound to minimum distance of:

$$d_{\min} \leq n - k + 1 \tag{8.9}$$

The only binary codes that achieve this bound with equality are simple $(n, 1)$ repetition codes; other upper bounds are usually tighter for binary codes. On the other hand Reed Solomon codes, which are multilevel codes, do have a minimum distance which is the maximum permitted by this bound. Reed Solomon codes were treated in Chapter 7.

8.6   GILBERT–VARSHARMOV BOUND

The Gilbert–Varsharmov bound shows that for a given value of n and k a certain value of minimum distance should be achievable by a linear block code. It does not necessarily mean that the code or codes which achieve this distance are known or have practicable implementations, merely that they exist.

Consider a code which has minimum distance d. The syndrome of an error pattern containing $d - 1$ errors may be the same as that of a single error, but no syndrome of an error pattern of weight $d - 2$ or less may be the same as a single-error syndrome. From this observation, we look at how we can make up the columns of the parity check matrix, which are just the syndromes of single-symbol errors, such that no column can be made from linear combinations of $d - 2$ or fewer other columns.

Each column of the parity check matrix contains $n - k$ symbols and for a q-ary code there are q^{n-k} possible columns. As we make up the columns, certain values are not allowed if we are to ensure that the column we are creating cannot be made from linear combinations of up to $d - 2$ previous columns. The problem of finding suitable columns becomes more acute as we fill the matrix and the last column (the nth) will be the most difficult. At this stage the prohibited combinations will be:

- All-zeros (1 possibility).

- Any of the $q - 1$ nonzero multiples of any of the $n - 1$ previous columns $[(n - 1)(q - 1)$ possibilities].

- A linear combination of nonzero multiples of i of the previous $n - 1$ columns, i.e. $[(n - 1)/i](q - 1)^i$ possibilities for each value of i from 2 to $d - 2$.

Hence we obtain

$$\sum_{i=0}^{d-2} \begin{bmatrix} n-1 \\ i \end{bmatrix} (q-1)^i < q^{n-k} \tag{8.10}$$

We are, however, allowed to choose values which are linear combinations of the possibilities up to $d-1$ of the previous $n-1$ columns, which gives us the full form of the Gilbert–Varsharmov bound:

$$\sum_{i=0}^{d-2} \begin{bmatrix} n-1 \\ i \end{bmatrix} (q-1)^i < q^{n-k} \le \sum_{i=0}^{d-1} \begin{bmatrix} n-1 \\ i \end{bmatrix} (q-1)^i \tag{8.11}$$

In this form, one can use the bound either to determine the maximum value of minimum distance that is sure to be available with a given set of q, n and k, or to set an upper limit to the value of $n-k$ that is needed to provide a desired minimum distance with a q-ary code of length n.

8.7   ERROR DETECTION

It has so far been assumed that the purpose of coding was to allow the receiver to recover the information from the received sequence with a higher certainty than would be obtainable without coding. Many error control schemes do not, however, attempt to recover the information when errors have occurred, instead they detect errors and invoke some alternative strategy to deal with them. If a return channel is available, the receiver may call for a retransmission of the message. Alternatively for data with considerable inherent redundancy it may be possible to reconstitute the corrupted message in a way that minimizes the effects of the loss of information.

There are many reasons why a system designer might opt for an error detection strategy rather than forward error correction. Some of these reasons are bound up with characteristics of an error detection scheme which will emerge in the course of this chapter. One major reason is, however, that error detection can be made many orders of magnitude more reliable than forward error correction and is thus appropriate when a low undetected error rate is essential. It is also often relatively simple to implement an error control strategy based on error detection. Thus if the characteristics are acceptable, error detection strategies, or some hybrid of error detection and forward error correction, are likely to be the most cost-effective solution.

8.8   RANDOM-ERROR DETECTION PERFORMANCE OF BLOCK CODES

In comparison with error correction, error detection is a relatively straightforward operation, but it is rather more difficult to obtain approximate formulas for the performance because the structure of the code has a much more noticeable effect. It is almost always block codes that are used and, although convolutional codes are possible, we shall look only at the performance of block codes.

If the number of errors in a block is less than the minimum distance then they will always be detected. If the number is equal to or greater than d_{min} then we might choose to be pessimistic and assume that error detection will fail. This, however, is far too removed from real performance; only a small proportion of error patterns of weight d_{min} or more will produce another codeword and hence escape detection. Taking the example (7, 4) code from Chapter 3, we see that there are seven codewords of weight 3, seven of weight 4 and one of weight 7. If therefore the all-zero codeword is transmitted, only seven of the 35 possible 3-bit error patterns produce an undetected error, so that 80% of 3-bit errors will be detected. The distance properties of the code are the same regardless of which codeword is transmitted, so this result applies to any transmission. Similarly 80% of weight 4 error patterns are detected, 100% of weight 5 and 100% of weight 6. Only the weight 7 error pattern is sure to evade detection.

Ideally we would wish to know the number A_i of codewords of weight i for the code in use. If we assume that the events causing code failure are essentially independent we can then say

$$P_{ud} = \sum_{i=0}^{n} P_i \frac{A_i}{\left[\begin{array}{c} n \\ i \end{array}\right]} \tag{8.12}$$

where P_{ud} is the probability of undetected error and $P(i)$ is the probability of exactly i symbols in a block being wrong. With a symbol error rate of p_s, we can see that

$$P_{ud} = \sum_{i=0}^{n} A_i p_s^{i} (1 - p_s)^{n-1} \tag{8.13}$$

Unfortunately the weight structures are not known for all codes. Nevertheless the weight distributions are known for Hamming codes, Reed Solomon codes and some binary BCH codes. In addition the weight distribution can be obtained for any code where the weight distribution of its dual code is known.

8.9   WEIGHT DISTRIBUTIONS

Hamming codes

Hamming codes have a *weight enumerator*

$$A(x) = \sum_{i=0}^{n} A_i x^i = \frac{(1+x)^n + n(1+x)^{(n-1)/2}(1-x)^{(n+1)/2}}{n+1} \tag{8.14}$$

that is, the coefficient of x^i in $A(x)$ is the number A_i of codewords of weight i. An alternative form is

$$A(x) = \sum_{i=0}^{n} A_i x^i = \frac{(1+x)^n + n(1-x)(1-x^2)^{(n-1)/2}}{n+1}$$

from which we can obtain expressions for A_i

$$A_i = \begin{cases} \dfrac{\begin{bmatrix} n \\ i \end{bmatrix} + n(-1)^{i/2} \begin{bmatrix} (n-1)/2 \\ (i-1)/2 \end{bmatrix}}{n+1} & i \text{ even} \\[4mm] \dfrac{\begin{bmatrix} n \\ i \end{bmatrix} + n(-1)^{(i+1)/2} \begin{bmatrix} (n-1)/2 \\ i/2 \end{bmatrix}}{n+1} & i \text{ odd} \end{cases} \tag{8.15}$$

For the (7, 4) Hamming code, $A_0 = 1$, $A_3 = 7$, $A_4 = 7$, $A_7 = 1$ and all the other terms are zero. This corresponds with the results quoted in the above section.

Reed Solomon codes

The weight distribution of a t-error correcting Reed Solomon code over GF(q) is given by $A_0 = 1$ and

$$A_i = \begin{bmatrix} q-1 \\ i \end{bmatrix}(q-1) \sum_{j=0}^{i-2t-1} (-1)^j \begin{bmatrix} i-1 \\ j \end{bmatrix} q^{i-2t-1-j} \tag{8.16}$$

for $2t + 1 \le i \le n$. An alternative (equivalent) form is

$$A_i = \begin{bmatrix} q-1 \\ i \end{bmatrix} \sum_{j=0}^{i-2t-1} (-1)^j \begin{bmatrix} i \\ j \end{bmatrix}(q^{i-2t-1-j} - 1) \tag{8.17}$$

For example a double-error correcting Reed Solomon code over GF(8) has 1 code-word of weight zero, 147 of weight 5, 147 of weight 6 and 217 of weight 7.

Dual of code of known weight distribution

For any (n, k) code it is possible to construct the $(n, n-k)$ *dual code* whose generator matrix is the parity check matrix of the original code. If the original code is a cyclic code with a generator $g(X)$, the dual code has generator $[X^n + 1]/g(X)$. The weight enumerator $A(x)$ of a (n, k) linear code over GF (q) is related to the weight enumerator $B(x)$ of its dual by the *MacWilliams Identity*:

$$q^k B(x) = [1 + (q-1)x]^n A\left[\frac{1-x}{1+(q-1)x}\right] \tag{8.18}$$

For binary codes this becomes

$$2^k B(x) = (1 + x)^n A\left(\frac{1 - x}{1 + x}\right) \tag{8.19}$$

For a Hamming code with weight distribution given by Equation (8.14), the MacWilliams identity gives the following expression for $B(x)$, the weight distribution of the dual code:

$$B(x) = 1 + nx^{(n+1)/2}$$

The dual of a Hamming code is in fact a maximal length code or simplex code. That this is indeed the weight distribution of such a code will be seen in Chapter 9 (Section 9.3).

If we know only the numerical values of the coefficients A_i instead of an analytic expression, we can still obtain the weight distribution of the dual code. For example if we take the values of A_i for the (7, 4) Hamming code we find from Equation (8.18)

$$16B(x) = (1 + x)^7 \left[1 + 7\left(\frac{1 - x}{1 + x}\right)^3 + 7\left(\frac{1 - x}{1 + x}\right)^4 + \left(\frac{1 - x}{1 + x}\right)^7\right]$$

$$16B(x) = (1 + x)^7 + 7(1 - x)^3(1 + x)^4 + 7(1 - x)^4(1 + x)^3 + (1 - x)^7$$

Expanding this gives

$$B(x) = 1 + 7x^4$$

The importance of the MacWilliams identity is that for a high rate code it is often much easier to find the weight distribution of the dual code, which will have far fewer codewords. In practice, therefore, the weight distribution of a Hamming code would be obtained from that of a simplex code, rather than vice versa as done here.

8.10 WORST CASE UNDETECTED ERROR RATE

Another possibility of interest is to consider as a worst case that the bit error rate approaches 0.5 when using a binary code. The probability of undetected error becomes

$$P_{ud} = \sum_{i=0}^{n} A_i 0.5^i (1 - 0.5)^{n-i}$$

but $\sum_{i=0}^{n} A_i = 2^k$, so

$$P_{ud} = 1/2^{n-k}$$

What this means is that if the bits are generated randomly then there is a chance of 1 in 2^{n-k} of the $n - k$ parity bits being correct. This is true only if the checks can be regarded as independent, and there are some codes where this is not so. Nevertheless the worst case probability of undetected error for well designed codes can be calculated in this way.

8.11 BURST-ERROR DETECTION

Cyclic codes have good burst-error detection properties. Any consecutive $n - k$ bits can act as the parity checks on the rest of the codeword and it therefore follows that an error pattern must span more than this number of bits if it is to pass undetected. The only bursts of length $n - k + 1$ which will pass undetected are those which are identical to the generator sequence cyclically shifted to the appropriate position. Thus over any fixed span of $n - k$ bits, there are 2^{n-k-1} error patterns starting and ending in 1, of which only one will pass undetected. Thus the probability of a burst of length $n - k + 1$ being undetected is $2^{-(n-k-1)}$.

This analysis extends fairly easily to longer bursts. For any burst of length $l > n - k + 1$ to pass undetected, it must resemble $g(X)$ multiplied by some polynomial of degree $l-(n-k)$. There are $2^{l-(n-k)-2}$ such polynomials and 2^{l-2} burst patterns of length l. Thus the probability of such a burst being undetected is $2^{-(n-k)}$.

8.12 EXAMPLES OF ERROR DETECTION CODES

There are three cyclic block codes which have been used frequently in error detection applications, for example in network protocols. One is a 12-bit cyclic redundancy check and the other two are 16-bit checks.

The generator polynomial for the 12-bit CRC is

$$g(X) = X^{12} + X^{11} + X^3 + X^2 + X + 1$$

or

$$g(X) = (X^{11} + X^2 + 1)(X + 1)$$

The polynomial $X^{11} + X^2 + 1$ is primitive; hence, the code is an expurgated Hamming code. The length of the code is 2047 ($2^{11} - 1$) of which 2035 are devoted to information and the minimum distance is 4. The code may be shortened to include less information without impairment to the error detection properties.

There are clearly too many codewords to enumerate fully the weight structure. Taking the codewords of weight equal to d_{min} we find that there are 4 44 34 005 codewords of weight 4 compared with 4.53×10^{10} possible weight 4 sequences. The probability of a weight 4 error sequence being undetected is therefore less than 10^{-3}. The code will detect all errors of weight less than 4, all errors of odd weight, all bursts of length less than 12, 99.9% of all bursts of length 12 and 99.5% of all bursts of length greater than 12.

The two 16-bit CRCs have generator polynomials

$$g(X) = X^{16} + X^{15} + X^2 + 1$$

and

$$g(X) = X^{16} + X^{12} + X^5 + 1$$

The factor $X + 1$ can be taken out to give

$$g(X) = (X^{15} + X + 1)(X + 1)$$

and

$$g(X) = (X^{15} + X^{14} + X^{13} + X^{12} + X^4 + X^3 + X^2 + X + 1)(X + 1)$$

In both cases the generator is a primitive polynomial multiplied by $X + 1$ to expurgate the code. As a result the codes have $d_{min} = 4$, length up to 65 535 of which all but 16 bits are devoted to information. There are 1.17×10^{13} words of weight 4, giving a probability of undetected error for weight 4 patterns of around 1.53×10^{-5}. The codes will detect all errors of weight 3 or less, all odd-weight errors, all bursts of length 16 or less, 99.997% of bursts of length 17 and 99.9985% of bursts of length 18 or more.

The data blocks on CD-ROM include a 32-bit CRC. In this case the polynomial consists of the product of two polynomials

$$g(X) = (X^{16} + X^{15} + X^2 + 1)(X^{16} + X^2 + X + 1)$$

The first of these factors is the first of the standard CRC-16 polynomials above. The second decomposes as

$$(X + 1)(X^{15} + X^{14} + X^{13} + X^{12} + X^{11} + X^{10} + X^9 + X^8 + X^7 + X^6$$
$$+ X^5 + X^4 + X^3 + X^2 + 1)$$

8.13   OUTPUT ERROR RATES USING BLOCK CODES

Suppose we are using a t-error correcting code and subjecting the decoder to a random bit error rate p. If we wish to work out the rate at which decoding errors occur, we usually assume that if more than t errors occur there will be a decoding error. This is an oversimplification because in general there will be some possibility of decoding beyond the guaranteed correction capability of the code or of detecting errors in such cases; it is therefore a pessimistic assumption. The use of the Hamming bound to estimate probability of detecting uncorrectable errors was discussed in Section 8.2.

If we use a binary code then the code symbol error rate p_s will be the same as the channel bit error rate p. On the other hand, suppose our code uses symbols which consist of l bits. The symbol error rate over a memoryless binary channel will be given by

$$1 - p_s = (1 - p)^l$$

In other words, the probability of a symbol being correct is equal to the probability of all the bits being correct. Rearranging this gives

$$p_s = 1 - (1 - p)^l \tag{8.20}$$

The probability $P(i)$ of i symbol errors out of n symbols is

$$P(i) = \begin{bmatrix} n \\ i \end{bmatrix} p_s^i (1 - p_s)^{n-i} \tag{8.21}$$

where

$$\begin{bmatrix} n \\ i \end{bmatrix} = \frac{n!}{i!(n-i)!}$$

The probability P_{de} of a block decoding error is just the sum of $P(i)$ for all values of i greater than t

$$P_{de} = \sum_{i=t+1}^{n} P(i) = 1 - \sum_{i=1}^{t} P(i) \tag{8.22}$$

If a decoding error occurs then we must decide whether we are interested in message or bit error rates. Message error rates P_{me} are the easier to calculate; if the message consists of m blocks then it will be correct only if all m blocks are correct:

$$1 - P_{me} = (1 - P_{de})^m$$

$$P_{me} = 1 - (1 - P_{de})^m \tag{8.23}$$

If we want the output bit error rate then we need to make a further assumption about the number $n_e(i)$ of additional errors output in an n-bit block when there is a decoding failure. At best the decoder will output d_{min} symbol errors, i.e. $d_{min} - i$ additional errors. At worst it will attempt to correct t errors and in so doing will add an extra t errors to those at the input

$$d_{min} \le n_e(i) \le i + t \tag{8.24}$$

In practice the statistics will be dominated by the case when $i = t + 1$ and the upper and lower bounds on $n_e(i)$ are identical. Only at fairly high input bit error rates will there be an appreciable difference between using the upper and lower bounds.

Having chosen a suitable expression for $n_e(i)$, the output symbol error rate P_{se} will be

$$P_{se} = \frac{1}{n} \sum_{i=t+1}^{n} n_e(i) P(i) \qquad (8.25)$$

and in the simplest case where we assume that $n_e = d_{min}$ the expression becomes

$$P_{se} \approx \frac{d_{min}}{n} \sum_{i=t+1}^{n} P(i) \qquad (8.26)$$

If the code is binary then this is the same as the output bit error rate p_0.

If the code uses l-bit symbols (e.g. a Reed Solomon code), then obtaining the outpit bit error rate is not straightforward. Assuming time domain encoding, we can look at all possible symbol values to see that on average 50% of the bits would be wrong, a total of 2^{l-1} bit errors. One symbol value, however, does not represent an error, leaving $2^l - 1$ possible symbol error patterns over which to average. This, however, would give the bit error rate only for the additional symbol errors created by the decoder. For the symbol errors created on the channel the bit error rate is p/p_s. The overall output bit error rate is therefore

$$p_0 = \frac{1}{n} \sum_{i=t+1}^{n} P(i) \left\{ \frac{ip}{p_s} + [n_e(i) - i] \frac{2^{l-1}}{2^l - 1} \right\} \qquad (8.27)$$

As seen in Chapter 7, RS codes can be encoded from the frequency domain. In that case, decoding errors in the time domain can be considered to have a random effect on the frequency domain information, resulting in an output BER of 50% in the incorrectly decoded sequences:

$$p_0 = 0.5 \sum_{i=t+1}^{n} P(i) \qquad (8.28)$$

8.14 DETECTED UNCORRECTABLE ERRORS

The expressions in the previous section rest on the assumption that the occurrence of detected uncorrectable errors is relatively uncommon. This assumption is not valid for long codes with large values of minimum distance. To see the problem and the possible ways of tackling it, we consider a particular example of a triple-error correcting (31, 16) binary BCH code. We can plot the probability of not obtaining correct output against E_b/N_0 and compare it with the probability of error in a 16-bit uncoded block as in Figure 8.1.

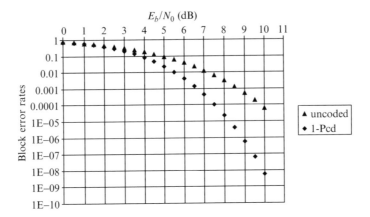

Figure 8.1 Block error rates for (31, 15) binary BCH code

The number of syndromes for this code (including the all-zero syndrome) is 2^{15}, i.e. 32 768. The total number of errors of weight ≤ 3 is

$$1 + 31 + \frac{31 \times 30}{2} + \frac{31 \times 30 \times 29}{3 \times 2} = 4992$$

Therefore the number of syndromes corresponding to correctable errors is only just over 15 % of the total. This figure (or one derived from a more accurate calculation) could be used to partition the decoded block error rates into incorrectly decoded blocks and detected uncorrectable errors.

As seen above, when errors of weight 4 or more occur we would, at worst, expect a 15 % chance that incorrect decoding will result. More often than not, the syndrome will correspond to an uncorrectable error. In such a case, the usual assumption for BER calculations is that the decoder will strip off the parity checks and output the information bits as received. Most blocks that cannot be decoded correctly will have four bit errors and the proportion of bit errors in the output will be 4/31 if the errors are detected as uncorrectable, compared with 7/31 if the block is miscorrected. Based on the usual assumption, the detection of errors will therefore reduce the output bit error rate by less than a factor of 2. This will have only a small effect on coding gain as shown in Figure 8.2, which assumes that all uncorrectable errors are detected.

As we can see, even if we assume that all the uncorrectable errors are detected, it makes little difference to the coding gain. This seems to conflict with common sense because the knowledge that a block is in error can be used to invoke some other strategy. In a selective repeat ARQ scheme, for example, the efficiency is considered to be the product of the code rate and the proportion of frames received correctly. It might therefore be better to consider that blocks with detected uncorrectable errors effectively reduce the code rate and increase the value of E_b/N_0. The result of that basis of assessment is shown in Figure 8.3. Comparison with Figure 8.2 shows that coding gain is increased by around 0.5 dB at a BER of 10^{-5}. The difference between this and the conventional basis of assessment will be even more significant for many more complex codes.

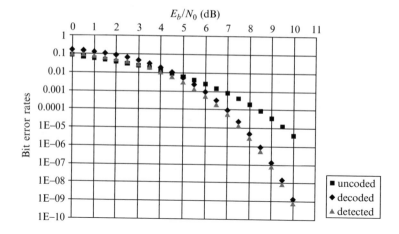

Figure 8.2   Bit error rates of (31, 16) code with and without error detection adjustment

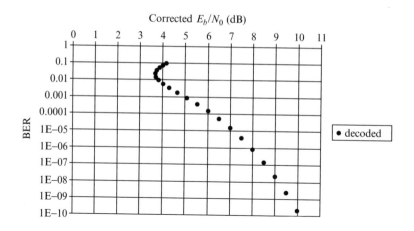

Figure 8.3   Performance of (31, 16) code with E_b/N_0 correction for detected errors

8.15   APPLICATION EXAMPLE – OPTICAL COMMUNICATIONS

Simple block codes are rarely favoured for communications applications, however there is one area – optical networks – in which they are universally used. Three factors conspire to make this so. Firstly the target BER is 10^{-15}, much lower than is encountered in radio or satellite communications. Secondly the bit rates are so high, up to 10G bits per second currently and evolving towards 40G, that implementing soft-decision demodulation is almost impossible and decoder complexity is also a significant issue. Finally, higher bit rates bring dispersion losses as well as increased noise through the increase in bandwidth. As a result high rate codes are needed.

The code that has been commonly adopted as a standard is a (255, 239) Reed Solomon code. The (conventionally defined) performance of this code is shown in Figure 8.4. It can be seen that the gain of this code at the target BER is a little over 6 dB. It is worth noting, however, that an equivalent length binary BCH code would give better performance. A (255, 239) binary BCH code would correct only two errors and the performance would be poor; however, the length of the RS code in bits is 2020, so a length 2047 binary BCH code would be comparable. A (2047, 1926) code is of slightly higher rate and corrects up to 11 bit errors. The performance is shown in Figure 8.5 and the gain can be seen to be improved by more than 0.5 dB relative to the RS code.

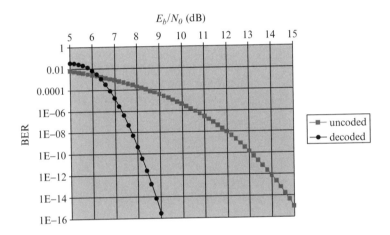

Figure 8.4 Performance of (255, 239) RS code

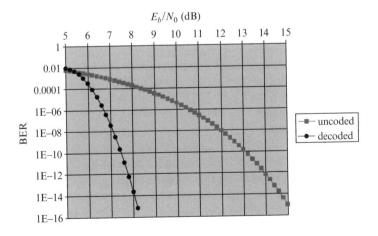

Figure 8.5 Performance of (2047, 1926) binary BCH code

Although ascertaining the reason for adoption of the Reed Solomon code is not easy, it is clearly not related to AWGN performance. Neither is it likely to be based on burst-error correction capability because the AWGN model is a good representation of the noise in the detectors, although it may be that there is a comfort factor in using RS codes in case burst errors happen to occur. The principal benefit of the RS codes is one of complexity; some important parts of the decoding algorithms have complexity that is proportional to length in symbols, and the binary BCH code would have been more difficult to implement. Even so, the need for higher gains for long haul and higher bit rate applications is expected to produce solutions that do not rely exclusively on Reed Solomon codes.

8.16 CONCLUSION

This chapter has looked at aspects of performance of linear block codes of all types, including what is theoretically possible, the performance of real examples and the applicability based on performance. The most commonly applied block codes are Reed Solomon codes, often in a multistage configuration. This will be the topic of the next chapter and several applications will be discussed there.

8.17 EXERCISES

1   How many codewords of weight 3 and weight 4 are there in a (15, 11) Hamming code? If the code were

   (a) expurgated to (15, 10) by removal of all odd weight codewords
   (b) expanded to (16, 11) by inclusion of an overall parity check

   how would these values change?

2   For a (15, 11) Hamming code, find the probability of undetected error for the following cases

   (a) random errors weight 3
   (b) random errors weight 4
   (c) burst errors length 4
   (d) burst errors length 5
   (e) burst errors length 6

3   Find the number of codewords of weight 5 and 6 in a (15, 11) Reed Solomon code. Hence find the probabilities that random errors affecting 5 or 6 symbols will be undetected.

4 An error detection scheme requires the worst-case probability of undetected errors to be 10^{-6}. How many parity bits are required?

5 Estimate the output bit error probabilities for Hamming codes of length 7, 15 and 32 in the presence of channel bit errors with probability 10^{-2}. Find the asymptotic coding gain of these codes and comment on the comparison between the different codes.

6 Estimate the probability of undetected decoding errors for a (31, 15) Reed Solomon code over GF(32).

9

Multistage coding

9.1 INTRODUCTION

There are many good ways in which coding may be applied in multiple stages. The intention is to create codes that are effectively long but whose decoding can be achieved by relatively simple component decoders. In principle multistage decoding is not optimum, however we shall see that in some cases the codes can be used in conjunction with feasible multistage decoding to give excellent performance. In Chapter 10 we shall see some cases where we can achieve performance approaching the Shannon limits.

9.2 SERIAL CONCATENATION

A common use for Reed Solomon codes is in serially concatenated coding systems where two codes are applied in sequence as shown in Figure 9.1. The first code to be applied is called the outer code and will in this case be a RS code. The second code is the inner code and is designed to work well for the channel conditions.

When the inner decoder fails, it produces a burst of errors. Since the Reed Solomon code is good at correcting bursty errors, these should be corrected by the RS decoder. The combination of inner code and channel produces a *superchannel*

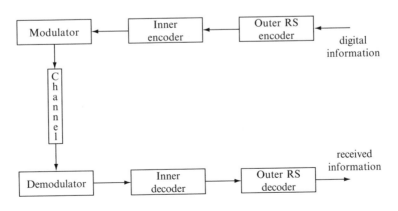

Figure 9.1 Concatenated code

that has, it is to be hoped, a lower error rate than the uncoded channel, and the superchannel errors are confined to bursts of known duration. The symbol size of the outer code is chosen such that inner decoding errors affect relatively few symbols, perhaps only one.

Serial concatenation can be thought of as a way of converting channel errors into bursts so that they can be tackled by a burst-error correcting code. It should be remembered that, for the same bit error rate, bursty errors are in principle easier to combat than random errors. It will also be seen that in some cases the burst errors can be constrained to be phased bursts corresponding with the natural boundaries of the Reed Solomon outer code. In this way the concatenation can make the most efficient use of Reed Solomon codes.

Producing reasonable concatenated schemes is actually a difficult problem, especially because of the need to balance the relative error-correcting power of the two codes. For a fixed overall code rate, it might be possible to reduce the rate and increase the power of one code, compensating with an increased rate and reduced error-correcting power of the other. Somewhere will be an optimum balance that is not easy to find. The inner code must be able to achieve reasonable error control on the channel and may need to be burst correcting or interleaved if the channel characteristics are predominantly bursty. If the inner code is not powerful enough, or not well suited to the channel, it is wasted and serves only to increase the error rates that the outer code has to handle. In such circumstances the outer code might do better on its own. Similarly the outer code must be able to cope with the symbol error rates on the superchannel.

9.3   SERIAL CONCATENATION USING INNER BLOCK CODE

In principle, any short block code could be used as the inner code. Choosing the dimension of the inner code to be equal to the symbol size of the outer code ensures that the superchannel errors are in phase with the symbol boundaries of the outer code for maximum efficiency. It does however restrict the values of k we would wish to use because of the rising complexity of Reed Solomon decoding for large symbols. On the other hand, short codes often give the realistic possibility of soft-decision decoding to provide a worthwhile extra gain.

There are certain specific families of block codes that have commonly been proposed and evaluated for use as inner codes. These are principally the maximal length (simplex) codes, orthogonal codes and biorthogonal (Reed Muller) codes which can be soft-decision decoded by correlative methods, giving optimum or near-optimum performance. The drawback to all these codes, however, is that they are very low rate, so some higher rate possibilities should be considered too.

9.3.1   Maximal length codes

Maximal length codes are codes which have length $n = 2^k - 1$ for dimension k. They are the *dual codes* of Hamming codes, which means that the generator matrix of a

Hamming code can be used as the parity check matrix of a maximal length code, and vice versa. They can be generated as cyclic codes by taking

$$g(X) = \frac{X^n + 1}{p(X)}$$

where $p(X)$ is a primitive polynomial of degree k.

The codewords consist of the all-zero sequence and the n cyclically shifted positions of the generator sequence. Thus there are $n + 1$ codewords, giving the relation between n and k shown above.

The minimum distance of a maximal length code is $2^k - 1$ and all the nonzero codewords are at this distance from the all-zero codeword. Linearity, however, means that the distance structure of the code looks the same from any codeword, so we reach the conclusion that every codeword is at the same distance from every other.

For a fixed outer code with symbol size k, the appropriate maximal length code as an inner code will allow operation at the lowest possible value of E_b/N_0.

9.3.2   Orthogonal codes

Orthogonal signals form a commonly used technique which may be regarded either as a modulation or as a low rate code. Viewed as a code, it is closely related to the maximal length code. An orthogonal signal set may be used as the inner code of a concatenated scheme.

Two signals $S_n(t)$ and $S_m(t)$ are orthogonal over some period T if

$$\int_0^T S_n S_m \, dt = \begin{cases} 0 & m \neq n \\ K & m = n \end{cases}$$

where K is some positive constant. Orthogonal signal sets are commonly provided, at the expense of bandwidth, by the use of M-ary frequency-shift keying (MFSK). Another way is to use an orthogonal code.

An orthogonal code results from adding an overall parity check to a maximal length code to produce a $(2^k, k)$ code. The code is less efficient than the maximal length code because the additional parity check is always zero and contributes nothing to the minimum distance of the code.

The code provides an orthogonal set of 2^k signals, one for each of the possible input values. A correlator looking for one codeword will, in the absence of noise, give a zero output if a different codeword is received. Provided there are not too many codewords, soft-decision decoding can therefore be achieved by a bank of correlators, each looking for one codeword, with the decoded value decided on the basis of the highest output. Note also that orthogonal symbols fit the GMD metric discussed in Chapter 7 (Section 7.10) and that the strength of the correlation could be used to determine inner decoder reliability. We might therefore expect useful performance gains from GMD decoding of the outer RS code used with an inner orthogonal code.

9.3.3   Reed Muller codes

Adding in the all-ones codeword to the generator matrix of an orthogonal code doubles the number of possible codewords and produces a biorthogonal code. The number of information bits has increased by one compared with the same length orthogonal code, but minimum distance is unchanged. Hence $n = 2^{k-1}$ and $d_{min} = 2^{k-2}$. Biorthogonal codes are also known as first-order Reed Muller codes.

The generator matrix of the first-order Reed Muller code has k rows and 2^{k-1} columns. Bearing in mind its derivation from maximal length codes, we can derive it in three stages. Firstly produce a $(k-1) \times (n-1)$ matrix whose columns consist of all the combinations of $k - 1$ bits except all-zeros. This is using the parity check matrix of a Hamming code as the generator of a maximal length code. Now add an all-zero column to represent the overall parity check of the orthogonal codes, producing a $(k-1) \times n$ matrix. Finally add another row which is all ones leaving a $k \times n$ matrix.

Rows of the generator matrices of higher-order Reed Muller codes can be produced by taking products of all pairs of the first $k - 1$ rows for second order, then all triplets for third order, etc. Higher rate codes are thus produced but the minimum distance is reduced by a factor of two for every increment in order.

All Reed Muller codes can be decoded by majority logic, although several steps are needed for the higher-order codes. Our interest here, however, is in the biorthogonal codes because, being so closely related to maximal length codes, they give very similar performance, and there are some implementation advantages. For small k, soft-decision decoding can be carried out using correlative techniques in which there are a number of matched filters looking for single codewords. Because half the codewords of the Reed Muller codes are the complements of the other half, we can use half the number of matched filters and take the sign of the correlation to decide which of the two codewords has been received.

9.3.4   High rate codes with soft-decision decoding

The codes of the above three sections have all been low rate codes, and their use in concatenated coding schemes reduces the rate still further. Other block codes are therefore often used to produce overall code rates of around 0.5 or greater. Soft-decision decoding is preferred for best overall gain, provided it is practicable.

There are several methods for soft-decision decoding of block codes that become practicable when the codes are short and which give a performance close to true maximum likelihood decoding. The Chase algorithm is probably the most widely used of these and is certainly applicable to concatenated schemes. We shall therefore study the method first before considering the codes to which it might be applied.

The intention of the Chase algorithm is to generate a list of codewords that will almost always contain the maximum likelihood codeword. The basic procedure consists of three steps. Firstly hard decisions are made on the bits of the received sequence. Secondly, a number of error sequences are generated and added to the hard-decision received sequence. Finally each of the sequences produced in step 2 is decoded and compared with the (soft-decision) received sequence, and the one which is closest is selected.

The important part of the algorithm is the method of generating the test patterns. If the hard-decision decoded result is not the best answer then one of its nearest neighbours usually will be, and the aim is that the error sequences should, after decoding, produce a good set of these near neighbours. Chase proposed three different methods which generate different numbers of error sequences, the largest number giving the best performance. However the 'second best' of his methods gives virtually identical performance to the best with greatly reduced complexity and is therefore the most widely encountered. The i least reliable bits are identified, where i is the largest integer which does not exceed $d_{\min}/2$. The error patterns consist of all the 2^i possible values in these positions and zero in the rest.

Typical short block codes used in concatenated schemes have $d_{\min}$ equal to 4. The Chase algorithm therefore requires only four decoding operations to achieve a performance that is, in typical operating conditions, within a few tenths of a dB of that obtained by maximum likelihood decoding.

Another alternative that is sometimes encountered is the use of a simple parity check code ($d_{\min} = 2$). To achieve maximum likelihood decoding, accept the hard-decision received sequence if the parity is satisfied, otherwise complement the least reliable bit.

9.4   SERIAL CONCATENATION USING INNER CONVOLUTIONAL CODE

When the inner code is a convolutional code, the choice of symbol size for the outer code is less straightforward than with an inner block code. We choose some multiple of k_0, usually at least $(m + 1)k_0$, because a decoding error will normally affect at least k_0 bits and the number of symbol errors will be reduced by having a large symbol. Standard schemes have used the rate $1/2$ $K = 7$ convolutional code and, because implementation of GF(128) RS codes is inconvenient, a code over GF(256) has been chosen. Unfortunately the errors may not correspond exactly with the symbol boundaries and we need to cater for one more symbol error than would be the case for phased errors. For example, phased burst errors of up to 32 bits long would affect only four RS symbols; however, non-phased bursts of length 26 to 33 could affect five symbols.

Whatever the choice of symbol size, it will usually be necessary to interleave several outer codewords together, symbol-by-symbol, in order to spread the inner decoding errors over several codewords. Decoding errors from a convolutional code typically last for a few constraint lengths and will therefore affect several symbols of the outer code. Interleaving will result in fewer occasions when the outer code is defeated because of the greater length over which inner decoding errors are averaged. Obviously the need is greater the smaller the symbol size of the outer code, but in any case spreading the errors widely will reduce the severity of fluctuations in symbol error rates and reduce the incidence of outer decoding errors. In the above example, we would certainly want to interleave to degree at least 5.

9.5   PRODUCT CODES

If we were to take a (n_1, k_1) inner block code, symbol interleave it to some degree k_2 and then apply a (n_2, k_2) outer block code with the same symbol size, we would produce an arrangement known as a product code, illustrated in Figure 9.2. Assuming linear codes are used, the segment of the array bearing the label 'checks on checks' will be the same regardless of whether the row code or column code is applied first.

Product codes are a way of producing complex codes from simple components. For example if both the row and column codes are simple parity checks (single-error detecting), the overall code can correct single errors. Any single error will fail one row and one column parity check, and these failures can then be used to locate the error. In general, if the minimum distance of the row code is d_1, correcting t_1 errors, and of the column code is d_2, correcting t_2 errors then for the product code the minimum distance is $d_1 \cdot d_2$, correcting $2 \cdot t_1 \cdot t_2 + t_1 + t_2$ errors. Achieving this performance, however, is not necessarily straightforward as the strategy for row and column decoding may not be easy to define.

Take, for example, the product of two single-error correcting codes, which should correct four random errors. If those errors are arranged so that there are two errors in each of two rows and two columns then a simple strategy of alternately correcting rows and columns will only make things worse. The row correction will add an extra error into each of the affected rows and the column correction will then do the same into each of the affected columns. The product code will thus be wrongly decoded.

There is, in fact, an approach to decoding which will always find a codeword if one exists within $(d_1 d_2 - 1)/2$ of the received sequence. Assuming that the order of transmission is along the rows, the decoding method relies on having an error-correcting decoder for the rows and an error and erasure correcting decoder for the columns:

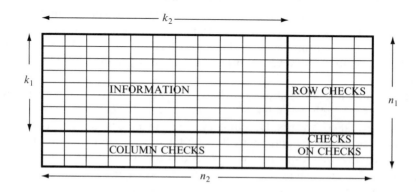

Figure 9.2   Product code

1   Decode the rows. For any row which cannot be decoded, erase it. For any row i in which corrections are made, record ω_i, the number of corrections.

2   If an odd number of row erasures has taken place, erase one more row choosing the one for which ω_i is largest.

3   Calculate the error correction capability of the product code as $[d_1(d_2 - e) - 1]/2$ where e is the number of rows erased.

4   Decode one column.

5   If decoding succeeds count $d_1 - \omega_i$ for every position in which the value is corrected and ω_i for every (unerased) position in which no correction occurs. If this count is less than or equal to the error correction capability of the code, then the column is correctly decoded. Otherwise, or if the original decoding failed, erase two more rows (with largest ω_i), recalculate the error correction capability, and repeat.

6   After each successful column decoding, move on to the next column. Rows previously erased remain erased.

There will often be a tie for which columns should be selected for erasure. Such ties may be broken arbitrarily.

Consider what happens when we have a product of single-error correcting codes and a 4-bit error pattern affecting two rows and two columns as above. Let the row and column codes be the (7, 4) cyclic Hamming code described in Chapter 4 and let rows 5 and 2 contain errors in bits 5 and 3, as shown in Figure 9.3. The row decoder now introduces errors into bit 2 of each of those rows and records that the rows have had single-error correction (Figure 9.4).

The column decoder first decodes column 6 with no error correction being needed. It counts the two row corrections that have taken place, compares with the error correction capability of the code (4), and accepts the column decoding. When column 5 is decoded, however, it introduces another error into bit 3 of the column. Since row 3 has had no errors corrected, accepting the column decoding implies that there were

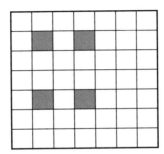

Figure 9.3   Quadruple error correcting product code with four errors

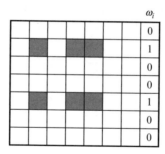

Figure 9.4   Product code after row decoding

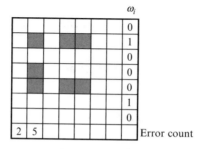

Figure 9.5   Product code after decoding of two columns

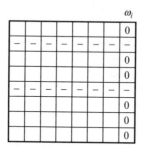

Figure 9.6   Product code after row erasure

three errors in that row in addition to the errors corrected in rows 5 and 2, making a total of 5 which exceeds the error correcting capability of the code (Figure 9.5). The column decoding is thus not accepted, rows 5 and 2 are erased (Figure 9.6), the error correction power of the code is now recalculated as $[3(3-2)-1]/2 = 1$, and the column decoder will now successfully fill the erasures.

Because of the interleaving, a product code will correct burst as well as random errors. Bursts of up to $n_2 \cdot t_1$ in length may be corrected if the order of transmission is along the rows. Moreover the code can correct bursts of this length or random errors at the same time. That does not mean that one array can contain a full complement of both burst and random errors, but that there can never be confusion between a burst error and a different random error pattern if both fall within the error correcting power of the code. The two patterns must have different syndromes and so can in principle be corrected by a single decoder. If the above decoding method were used with transmission down the columns then it is apparent that the effect of interleaving will be to ensure that burst errors are corrected as well as random errors. In fact this will also be the case if transmission is along the rows; miscorrected rows containing the burst will eventually be erased as the column decodings fail.

If n_1 and n_2 are relatively prime and the row and column codes are both cyclic, then the whole product code can be turned into a cyclic code provided an appropriate transmission order is chosen. The appropriate order is such that bit j in the transmitted stream is found in column $j \bmod n_1$ and in row $j \bmod n_2$, as shown in Figure 9.7. The generator polynomial of the cyclic product code is the highest common factor of $X^{n_1 n_2} + 1$ and $g_1(X^{bn_2})g_2(X^{an_1})$ where $g_1(X)$ and $g_2(X)$ are the generators of the row and column codes, respectively, and the integers a and b are chosen such that $an_1 + bn_2 = 1 \bmod n_1 n_2$.

The random and burst error correcting properties of the code are not diminished by ordering the transmission in this way, so an appropriate decoder for the cyclic product code should provide maximum decoding capability. However the purpose of product codes is to decode in relatively simple stages whilst still approaching optimum performance. The above properties are therefore of more theoretical than practical significance.

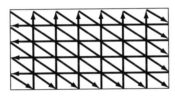

Figure 9.7   Cyclic ordering for product code

9.6   GENERALIZED ARRAY CODES

Product codes are only one way of producing a two-dimensional array coding scheme. Suitable modifications usually result in a code that is better in terms of the minimum distance achieved for a given code rate. Generalized array codes (GACs) allow us to create a variety of code parameters from simple components. The procedure for design is to create a product code, increase dimension by superimposing other codes on the parity checks in a way that does not affect the code properties, then finally remove symbols (if necessary) to reach the desired code length. It is found that many standard block codes can be created in this way.

Moreover, GACs possess interesting trellis structures that may allow low complexity soft-decision decoding.

Consider the example of creating a (7, 4) code with $d_{min} = 3$. We proceed as follows. We create a 2×4 array as shown in Figure 9.8, where three information bits i_2, i_1 and i_0 are given an overall parity check as a row code and the column code is a repetition code.

This is a (8, 3, 4) code. We now demonstrate that rate of the code can be increased. If we add another information bit into every location in the bottom row, either the extra bit is zero leaving the minimum weight unchanged, or else the bottom row is the one's complement of the top row resulting in codewords of weight 4. Therefore the code as shown in Figure 9.9 has minimum distance 4.

Finally, we remove the last parity check from the code to create the (7, 4, 3) code in Figure 9.10.

A trellis representation of the single-parity check code used as the basis of the construction is shown in Figure 9.11. The state denotes the running value of the parity bit as the information bits are generated. The labels on the trellis transitions indicate the code bits, the first three being the data and the last one being the generated parity check.

i_2	i_1	i_0	$i_2+i_1+i_0$
i_2	i_1	i_0	$i_2+i_1+i_0$

Figure 9.8   Initial product code

i_2	i_1	i_0	$i_2+i_1+i_0$
i_3+i_2	i_3+i_1	i_3+i_0	$i_3+i_2+i_1+i_0$

Figure 9.9   Augmented product code

i_2	i_1	i_0	$i_2+i_1+i_0$
i_3+i_2	i_3+i_1	i_3+i_0	

Figure 9.10   GAC representation of (7, 4, 3) code

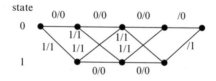

Figure 9.11   Trellis of single-parity check code

Taking into account the fact that the code is repeated on the bottom row gives a trellis as in Figure 9.12.

The (4, 1) repetition code used to augment the bottom row has a trellis shown in Figure 9.13.

The overall code trellis for the (8, 4, 4) code is shown in Figure 9.14. This is known as the Shannon product of the two trellises above and is constructed as follows.

First, obtain the state profile of the trellises, i.e. the number of states at each stage. From Figures 9.11 and 9.12 we have [2, 2, 2, 1] as the state profile for each trellis.

Second, obtain the branch profile, i.e. the number of branches at each stage. From Figures 9.11 and 9.12 we have [2, 4, 4, 2] and [2, 2, 2, 2], respectively.

Third, obtain the input label profile, i.e. the number of input symbols on the branches at each stage. From Figures 9.12 and 9.13 we have [1, 1, 1, 0] and [1, 0, 0, 0], respectively.

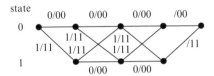

Figure 9.12   Trellis of repeated single-parity check code

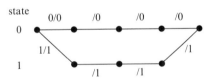

Figure 9.13   Trellis of repetition code

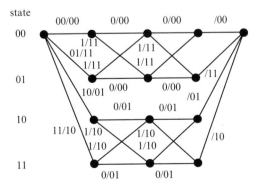

Figure 9.14   Trellis of (8, 4, 4) code

The Shannon product has at each stage a number of nodes which is the product of the number of nodes for each of the component trellises. The state number is the concatenation of the two states. If component trellis T_1 has a branch at depth i from state $S_{1,i}$ to state $S_{1,i+1}$ and component trellis T_2 has a branch at depth i from state $S_{2,i}$ to state $S_{2,i+1}$, the product trellis will have a branch from state $S_{1,i}S_{2,i}$ to state $S_{1,i+1}S_{2,i+1}$. The input label profile will be the sum of the corresponding figures for the two trellises.

We see that the state profile of the product is [4, 4, 4, 1], the branch profile is [4, 8, 8, 4] and the input label profile is [2, 1, 1, 0]. The lower trellis in the central section is just a duplication of the upper one with the second bit of each output pair (corresponding to the output bit from the repetition code) the inverted value of the first. To obtain the trellis for the (7, 4, 3) code, simply delete the second output label on the last frame, as shown in Figure 9.15.

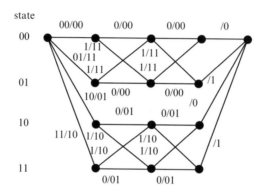

Figure 9.15   Trellis of (7, 4, 3) code

9.7   APPLICATIONS OF MULTISTAGE CODING

The first mass market application of error control coding was for the digital compact disc (CD). The information is recorded onto the disc using a run length limited (RLL) eight to fourteen (EFM) modulation code. Eight bits of information are mapped into a 14-bit sequence in a way that ensures the number of consecutive recorded bits having the same value is at least 3 and at most 10. The lower limit increases the recording density that can be used on the disc, whilst the upper limit reduces the possibility of synchronization problems. Three extra bits are added at the end of each 14-bit sequence to preserve the run length properties at the point where each sequence joins the next one and to improve the low-frequency suppression (for the benefit of the servo systems). Codes of this type [1] are regularly used to increase capacity of mass storage systems and are of considerable interest from the information theory point of view. However the main interest here is in the error control system which uses a

Figure 9.16   CD encoding

concatenation of two Reed Solomon codes in a scheme known as a cross-interleaved RS code (CIRC) [2].

The error control encoding scheme applied before the data is recorded to the disc is illustrated in Figure 9.16. The samples are 16-bit quantities and are separated into upper and lower bytes. The RS codes are over $GF(2^8)$, shortened to the lengths shown.

The D-interleave is a single delay to alternate bytes to ensure that errors affecting adjacent bytes on the channel end up in different codewords. The D^* interleave is a convolutional interleave to spread the errors from the (32, 28) decoder. The delays increase in steps of 4 to take account of the groupings of four bytes corresponding to left and right channels, high and low bytes of the sample. Each of these byte streams from the decoder ends up in a different codeword when the (28, 24) code is decoded. The Δ interleave is a novel feature intended for error concealment. It consists of a fixed reordering plus delays of 2 applied to half the bytes, again in groups of four. If there are detected uncorrectable errors from the (28, 24) decoder, the samples before and after those affected will fall in different codewords and are therefore likely to be correctly decoded. The erroneous sample values can thus be interpolated from the adjacent samples to reduce the likelihood of audible disturbance.

A major use of the CD format is for CD-ROM [3]. In data applications there is no inherent redundancy and the error concealment capability cannot be used. The protection afforded by the CIRC is not considered to be sufficient, and an additional product code is used prior to the CIRC. The convolutional interleaving is also a problem for CD-ROM as it effectively converts the code into a convolutional code, suitable for data in a continuous stream. There therefore needs to be a clearing process at the end of any data section.

The data on CD-ROM has a block structure in which 12 bytes of synchronization, a 4-byte header and 2048 bytes of user data forms the input to a product code using two parallel arrays, each of 43 columns and 24 rows. Each location holds a single byte, with one array holding all the MSBs and the other all the LSBs. The data is written into the array in row order and a (26, 24) RS code, known as the P code, is applied to the columns. The second code (the Q code) is a (45, 43) code applied to the diagonals of the array. If the diagonal reaches the bottom of the array (including the column parity checks), it continues from the top of the next column. A 4-byte parity check, 8 zero bytes, the 172 parity bytes from the P code (2 arrays × 43 columns × 2 parity checks) and 104 parity bytes from the Q code (2 arrays × 26 diagonals × 2 parity checks) form 288 bytes of auxiliary data to complete the block.

The digital versatile disc (DVD) [4] employs a more conventional Reed Solomon product code. The data is placed into an array of 172 bytes per row and 192 bytes per column. The row code is (182, 172) and the column code is (208, 192).

Concatenation of a Reed Solomon code with an inner convolutional code is a common configuration. The decoder for the rate $1/2$ $K = 7$ code, for example, will produce occasional errors that could easily affect 4 or 5 consecutive bytes and so some interleaving is used to spread these errors across several RS codewords. The interleaving, applied between the outer and inner codes, would be of symbols of the RS code; it would be disadvantageous to break up the bits of the individual symbols as this would increase the number of erroneous symbols.

Satellite communications applications have often used concatenations of Reed Solomon and convolutional codes. DVB(S) (Digital Video Broadcast by Satellite) is based around the MPEG-2 standard [5]. It uses a (204, 188) Reed Solomon code and a rate $1/2$ $K = 7$ convolutional code that can be punctured to rate $2/3$, $3/4$, $5/6$ or $7/8$. The rate used depends on the application and the bandwidth available. The interleaving between the outer and inner codes is convolutional of degree 12.

For deep space communications, NASA has a standard concatenation scheme employing a (255, 223) RS outer code, a rate $1/2$ inner convolutional code and block interleaving of degree of 4 between the inner and outer codes. This code can achieve a BER of 10^{-6} at E_b/N_0 just over 2.5 dB [6]. It was first used for the *Voyager* mission to send pictures from Uranus in January 1986. The European Space agency adopts a CCSDS standard that uses this code, but with interleaving degree up to 5.

9.8 CONCLUSION

The performance of conventional multistage coding, in particular serially concatenated codes, represented the closest approach to the Shannon limit until the advent of iterative decoding (Chapter 10) opened up other possibilities. Serial concatenation is a good way to create long codes with a feasible decoding algorithm because of the relatively simple components. Array codes also build long codes from simple components. There have therefore been many applications using these approaches. Generalized array codes and trellis approaches to block codes have been comprehensively described in [7].

9.9 EXERCISES

1   Messages of length 1100 bits are to be sent over a bursty channel. The bursts last for up to 10 bits. Suggest a concatenated scheme using a triple-error correcting outer code and a (15, 11) BCH inner code.

2   The scheme devised in question 1 is to be used on a Gaussian channel with bit error rate of 10^{-2}. Estimate the output bit error rate and the coding gain at this BER, assuming BPSK modulation.

3   Find the codewords of the cyclic code whose generator polynomial is $g(X) = X^4 + X^3 + X^2 + 1$. By inspection, verify that the code has the properties described for a maximal length code, i.e. all nonzero codewords being cyclic shifts of a single sequence.

The demodulator gives soft decisions with 3-bit quantization. The sequence 0 1 7 7 5 5 2 is received, where value 0 represents high confidence in received zero and value 7 represents high confidence in received 1. Values 0–3 represent hard-decision zero and values 4–7 represent hard-decision one. Find the closest code-word to the received sequence. Verify that the Chase algorithm, as described in Section 9.3.4, would generate the maximum likelihood codeword.

If the probability of decoding error from this code is 1%, suggest an outer code which will give a block decoding error rate below 10^{-4}. Find the overall code rate. Would interleaving the outer code produce a lower overall decoding error rate?

4   A rate 1/2 $K = 7$, convolutional code produces decoding errors which last for up to 30 frames. Suggest an outer code with minimum distance 11 and an appropriate interleaving degree.

5   A product code uses a (7, 4) Hamming code for the rows and a (15, 11) code for the columns. Find the minimum distance of the code and one codeword which has a weight equal to this minimum distance. Compare the burst-error correction capabilities if the code is transmitted across the rows or down the columns.

6   In the example of product code decoding, illustrated in Figures 9.3–9.6, it is found during the final column decodings that the product code is still capable of correcting one more error. Does this mean that a 5-error pattern could have been reliably decoded? What other significance could attach to the remaining error correction capability?

7   Construct a (24, 12) extended Golay code as a generalized array code.

9.10   REFERENCES

1   K.A. Schouhamer Immink, *Codes for Mass Data Storage Systems*, Shannon Foundation Publishers, The Netherlands, 1999.
2   IEC International Standard 908, *Compact disc digital audio system*, 1st edition, 1987.
3   ISO/IEC International Standard 10149, *Information technology – Data interchange on read-only 120 mm optical data disks (CD-ROM)*, 2nd edition, July 1995.
4   H-C. Chang and C. Shung, *A Reed-Solomon product code (RS-PC) decoder for DVD applications*, Proceedings of IEEE International Solid-State Circuits Conference '98, pp. 24.7-1–24.7-12, Toronto, February 1998.
5   ETSI, EN 300 421 'Digital Video Broadcasting (DVB): framing structure channel coding and modulation for 11/12 GHz satellite services', V1.1.2, August 1997.
6   J.H. Yuen (ed), *Deep Space Telecommunications Systems Engineering*, Plenum Press, 1983.
7   B. Honary and G. Markarian, *Trellis Decoding of Block Codes – A Practical Approach*, Kluwer Academic Publishers, 1997.

10
Iterative decoding

10.1 INTRODUCTION

In Chapter 9 we saw that product codes can be decoded by successive decoding of rows and columns, even though special measures may be needed to avoid getting stuck in suboptimum solutions. There are now many code constructions being proposed for telecommunication systems in which iterative soft-decision processing is used to obtain high coding gains. However, to carry out iterative soft-decision decoding, we need to have a decoding algorithm that can give soft outputs, because the output of one decoder becomes the input to another (soft-decision) decoder. This chapter will look at the algorithms available and the code configurations used for iterative soft-decision decoding.

10.2 THE BCJR ALGORITHM

The BCJR algorithm is a soft-in-soft-out algorithm named after its inventors, Bahl, Cocke, Jelinek and Raviv [1] who, in the same paper, were also the first to point out that a linear block code could be given a trellis representation. The principle of BCJR decoding is that the metric associated with a particular branch in the trellis is given by the product of probabilities associated with getting to the appropriate start node, following the branch and getting from the end node to the end of the trellis. We will use logarithmic metrics in the implementation, so that we will be adding the metrics rather than taking a product. To compute the SD output for each bit, we will compute the logarithmic metric for branches corresponding to a data 1 and subtract that for transitions corresponding to a data 0.

The metric used for BCJR decoding is defined as

$$\ln\left[\frac{p(1|r)}{p(0|r)}\right] \equiv \ln\left[\frac{p(r|1)}{p(r|0)}\right] + \ln\left[\frac{p(1)}{p(0)}\right] \tag{10.1}$$

where r is the received value of the bit. This metric will be in the range $-\infty$ to $+\infty$, positive for bit values tending to 1 and negative for bit values tending to 0. The term $\ln[p(1)/p(0)]$ represents *a priori* information that we may have about bit values. If there is none then this term becomes zero and the metric is effectively the log-likelihood ratio, as was used for Viterbi decoding. When decoding on a trellis, the branch metrics are computed from +bit metric for a transmitted 1 and −bit metric for a transmitted 0.

The first stage of decoding is similar to Viterbi decoding except that where two paths merge, instead of eliminating the less likely path, we compute a joint metric for arriving at that state by either route. The method of computing this joint probability will be given later. There is no need for path storage as no unique path is determined.

The second stage is like the first, except that the process is computed backwards from the end of the trellis.

To find data bit value for a given stage in the trellis:

(a) For each branch corresponding to a data 1, add the forward path metric to the start of the branch, the branch metric and the reverse path metric to the end of the branch.

(b) Compute a joint metric for the branches corresponding to a 1.

(c) Subtract the joint metric for the branches corresponding to a 0.

10.3 BCJR PRODUCT CODE EXAMPLE

Suppose we have a 3×3 product code in which each row and each column is a (2, 1) single parity check code. Suppose also that we receive the following values as in Figure 10.1.

We carry out a decoding of the rows using the BCJR algorithm, followed by a decoding of the columns. In the parlance of iterative decoding, row decoding followed by column decoding is called one *iteration* and a single set of decoding operations (row or column) is called a half iteration.

From the discussion in the previous chapter, Section 9.6, the single parity check code has the trellis as shown in Figure 10.2.

0.8	0.7	0.1
0.7	0.1	0.8
−0.9	0.9	0.8

Figure 10.1   Received values in product code

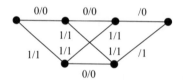

Figure 10.2   Trellis for (3, 2) code

To carry out the BCJR calculation for the top row of the code, first derive the branch metrics. These are shown in Figure 10.3.

To compute the path metrics, at stage 1 they are just the branch metrics but at stage 2 it is necessary to combine the metrics from two merging paths. If the two merging paths have metrics a and b, because these are logarithmic measures the way to combine them is to compute

$$\ln\left[\exp\left(a\right) + \exp\left(b\right)\right]$$

In our case, because of the symmetry of the trellis, we will always be combining metrics $+a$ and $-a$, in which case we compute $\ln\left[\exp\left(a\right) + \exp\left(-a\right)\right]$ $= a + \ln\left[1 + \exp\left(-2a\right)\right]$. The path metrics to stage 2 are the combination of $+1.5$ and $-1.5\ (= 1.55)$ and of $+0.1$ and $-0.1\ (= 0.70)$.

At stage 3 we would need to combine two nonsymmetric metrics; however, the result is not used in the final computation and so will not be calculated. The forward path metrics are therefore as shown in Figure 10.4.

The reverse path metrics are calculated in the same way, with the result shown in Figure 10.5.

As the final stage, we combine for all branches corresponding to a data bit 1, the sum of the forward path metric to the start of the branch, the branch metric and the

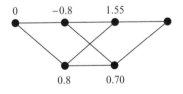

Figure 10.3   Example branch metrics

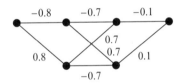

Figure 10.4   Forward path metrics

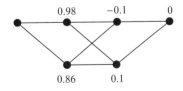

Figure 10.5   Reverse path metrics

reverse path metric to the end of the branch. We do the same for branches corresponding to data zero and subtract.

At stage 1 there is a single branch corresponding to data 1, the metric for which is $0.8 + 0.86 = 1.66$. The single branch corresponding to data 0 is $-0.8 + 0.98 = 0.18$. The metric for this bit is therefore 1.48.

At stage 2 there are two branches corresponding to data 1, with metrics $-0.8 + 0.7 + 0.1$ ($= 0.0$) and $0.8 + 0.7 - 0.1$ ($= 1.4$). The combination metric is 1.62. The two branches corresponding to data 0 have metrics $-0.8 - 0.7 - 0.1$ ($= -1.6$) and $0.8 - 0.7 + 0.1$ ($= 0.2$), for which the combination is 0.35. The final metric for this bit is therefore 1.27.

In the final frame there is no data, so we leave the received parity value unchanged at 0.1.

Carrying out the same process for the second row results in decoded values 1.27, -0.65 and 0.8. Carrying out the same process for the third row results in decoded values -2.83, 2.83 and 0.8.

We now have the array shown in Figure 10.6.

We now carry out the same decoding process for the columns and finish with the result shown in Figure 10.7. If we conclude the decoding at this point, the decoded information is 1 for a positive result and 0 for a negative result. We can see that the hard-decision-decoded values do satisfy the parity checks and so we expect no changes with further decoding.

1.48	1.27	0.1
1.27	−0.65	0.8
−2.83	2.83	0.8

Figure 10.6   Product code after row decoding

5.46	3.83	−0.75
5.44	−3.80	1.47
−2.83	2.83	0.8

Figure 10.7   Product code after row and column decoding

10.4   USE OF EXTRINSIC INFORMATION

When we carried out the column decoding, clearly we had some *a priori* information about the bit values as well as the received values. Given that the BCJR metric is the sum of received and *a priori* metrics, the *a priori* information going into the column decoder is the difference between the output and the input of the row decoder. A single component decoder therefore effectively behaves as shown in Figure 10.8. The information passed to the next component decoder is called the *extrinsic information* and is used as *a priori* information by the next decoder in conjunction with the received values.

Although this view of decoding did not affect the way the column decoding was implemented, in any situation where there is nonzero *a priori* information it clearly does affect the implementation. As the *a priori* information has already been used to derive the decoder output, failure to remove it before the next stage of decoding would effectively mean that it was doubly counted. Therefore, if we were to carry out further iterations of the above product code example, we would first need to derive the extrinsic information from the column decoder by subtracting the values in Figure 10.6 from those in Figure 10.7. The result is shown in Figure 10.9. Finally we would add the received values to create the sequence for the next iteration of the decoding, as shown in Figure 10.10.

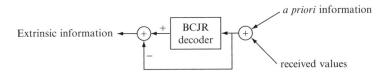

Figure 10.8   Component decoder schematic

3.98	2.56	−0.85
4.17	−3.15	0.67
0	0	0

Figure 10.9   Extrinsic information from column decoder

4.78	3.26	−0.75
4.87	−3.05	1.47
−0.9	0.9	0.8

Figure 10.10   Input to second iteration

10.5   RECURSIVE SYSTEMATIC CONVOLUTIONAL CODES

For many applications, the component code to be used with iterative decoding is a recursive systematic convolutional code (RSC code). These are used in a modified product code (different interleaving, no checks on checks), known as a *parallel concatenation*, to produce what is commonly called a *turbo code*. Codes of this type were first proposed by Berrou *et al.* [2, 3] and have led to a radical reappraisal of available performance from codes and an interest in iterative decoding applied to many types of combined processes in communications.

Strictly speaking, the word *turbo* applies to the iterative decoding process and so the term turbo code is something of a misnomer. More properly they should be referred to as turbo-decoded parallel concatenated RSC codes or iteratively decoded parallel concatenated RSC codes.

An example RSC encoder is illustrated in Figure 10.11.

The generator of the code is designated (1, 5/7). The 1 indicates that the data bit passes through into the code, the 5 indicates that the feedforward polynomial is $D^2 + 1$ (101 being the binary representation of 5) and the 7 indicates that the feedback polynomial is $D^2 + D + 1$ (111 being the binary representation of 7). Note however that some authors may show the polynomial coefficients in reverse order which will affect the representation of nonsymmetric polynomials.

If we create the state diagram of this code, the result is in Figure 10.12.

Comparison with the state diagram of the example code in Chapter 2 (Figure 2.2), which had generators (7, 5), shows that the actual code sequences of our RSC code are identical but that the mapping from information to code is different. The (7, 5) code without recursion has therefore become a (1, 5/7) code with recursion. This means of conversion from standard convolutional code with generators (g_1, g_2) to RSC code with generators $(1, g_1/g_2)$ is general.

To create a scheme for iterative decoding, the configuration is as in Figure 10.13. The information is generated for transmission and is also fed into an RSC encoder to generate a stream of parity check bits. At the same time the information is interleaved (denoted by Π) and fed into a second RSC encoder to generate a second stream of parity bits. The two encoders may be (indeed usually are) the same, but if the interleaving is good the two parity streams will be sufficiently independent to provide good performance with iterative decoding.

The code as illustrated is of overall rate 1/3, having one parity bit from each stream for each bit of information. It is possible, however, to puncture the parity

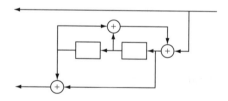

Figure 10.11   RSC encoder

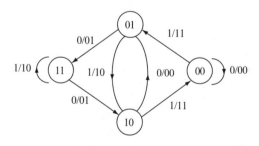

Figure 10.12   State diagram of (1, 5/7) RSC code

streams to increase the effective rate. For example, alternate bits could be taken from each of the two streams to produce a rate 1/2 code.

The decoder configuration is shown in Figure 10.14. Initially the *a priori* information is zero. From the first decoder the extrinsic information is derived and the received information added back in, then interleaving creates the correct ordering of information symbols for decoder 2. At the output of the second decoder, the extrinsic information is derived and deinterleaved before forming the *a priori* information for the first decoder on the next iteration.

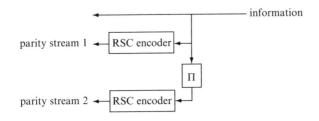

Figure 10.13   RSC encoding for iterative decoding

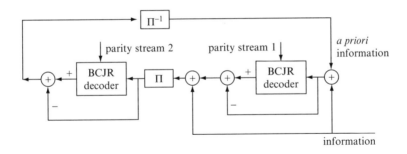

Figure 10.14   Iterative decoding for RSC codes

10.6   MAP DECODING OF RSC CODES

The BCJR algorithm is also known as maximum *a posteriori* (MAP) decoding, and is applied to the component RSC codes. We shall consider an example applied to the trellis for the (1, 5/7) code seen earlier. A transmission lasting 6 frames is considered and the encoder is assumed to be returned to state zero and the end of the example. The trellis is shown in Figure 10.15. The diagram takes the transmitted values as +1 (corresponding to bit value 1) or −1 (corresponding to bit value 0). The corresponding information values have not been shown.

Figure 10.16 shows the branch metrics for a received sequence 1.2, 0.8, 0.3, −0.9, −0.5, −0.8, −1.4, 1.1, −0.5, 1.3, 0.7, −0.2. Each bit metric is the product of the bit value on the trellis and the received value. Received values are assumed to be log likelihood ratios.

Figure 10.17 shows the forward path metrics for the example. In the early stages this is done by summing the branch metric and the path metric as usual. However, when two paths merge we need to compute the joint path metric. As before, if the two merging paths have metrics a and b, we compute $\ln[\exp(a) + \exp(b)]$. For example,

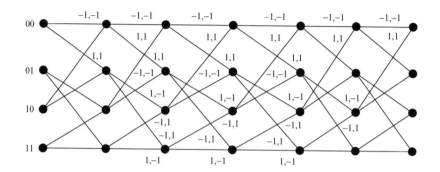

Figure 10.15 Trellis of RSC code

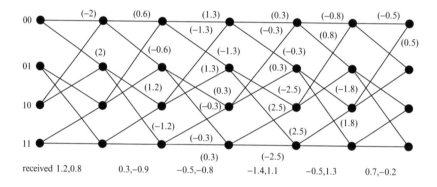

Figure 10.16 Example branch metrics

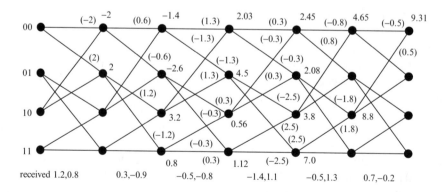

Figure 10.17 Example forward path metrics

at stage 3 the two paths into node 00 have metrics -0.1 and $+1.9$. The joint metric is $+2.03$ (slightly more than the larger value). The forward path metric for the last node has been shown but will not be needed.

Figure 10.18 shows the reverse path metrics. This repeats the previous process, but working backwards through the trellis. The reverse path metric to the start of the trellis is shown here but will not be needed.

Figure 10.19 shows the forward and reverse path metrics to every node and every branch metric (reverse path metrics underlined). We can now calculate the decoder output. Taking stage 2 as an example. There are two branches (00–01 and 01–10) corresponding to input 1. These have total metrics of $-2.0 - 0.6 + 2.61 = 0.01$ and $2.0 + 1.2 + 6.1 = 9.3$ respectively. The joint metric for these two branches is 9.3. There are two branches (00–00 and 01–11) corresponding to input 0. These have total metrics of $-2.0 + 0.6 + 3.56 = 2.16$ and $2.0 - 1.2 + 3.17 = 3.97$ respectively. The joint metric for these two branches is 4.12. The decoded output value is therefore 5.18. This corresponds to a data value 1.

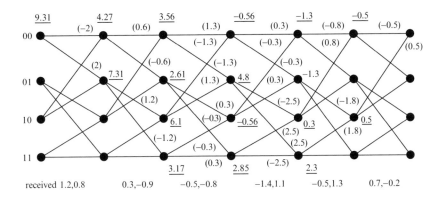

Figure 10.18   Example reverse path metrics

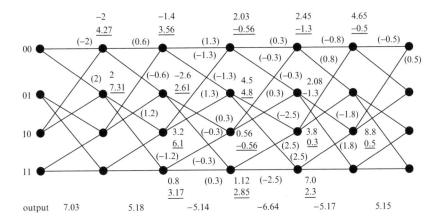

Figure 10.19   Decoder output calculation

The decoded (hard-decision) sequence is 1 1 0 0 0 1. From Figure 10.12 it can be seen that this does indeed correspond to a path terminating at state zero.

The version of the MAP algorithm shown here is technically known as *log-MAP* because it uses the addition of a logarithmic metric rather than multiplication of a probability ratio. However this creates difficulties in the combination of metrics for merging paths. Approximations can be used to simplify this. Note that if a is the larger metric, the combined metric can be approximated by $a + \exp(b - a)$. Another approximation is merely to take the larger value; this is known as *max-log-MAP* [4].

10.7   INTERLEAVING AND TRELLIS TERMINATION

The function of the interleaver is to decorrelate the inputs to the constituent encoders, so that the parity checks are independent. This improves the likelihood that an error pattern will eventually be correctable by one or other of the decoders. The interleaving usually employed is pseudorandom over a fixed-length block. In other words, the bits are written into a fixed-length buffer and then read out in some pseudorandom order that is known to both transmitter and receiver. Many other types of interleavers have also been designed, the intention being to produce the best performance within certain constraints of storage or of delay. Both block and convolutional interleaver designs are possible.

If the code is to be punctured to half rate, then it is desirable for the interleaver to have an *odd–even* property, i.e. that the odd- and even-position data bits are separately interleaved with the odd-position bits remaining in odd positions and the even-position bits remaining in even positions. Denoting the parity bits taken from the noninterleaved encoder as odd-position bits, these have been derived as odd-position information bits entered the encoder. If we want the parity checks from even-position information bits to be selected from the interleaved stream, then the odd–even property will ensure that this happens.

It is possible to avoid any need for termination of the trellis by using a *tail-biting* code. The encoding operation is started from the same trellis state in which it finishes at the end of the data sequence (or interleaved data sequence). This converts the trellis into a cylindrical structure (with the end wrapped round to the beginning) and the decoded path must be a continuous path around the cylinder. This approach has advantages of economy and avoidance of end effects, but difficulties in complexity of implementation. As a result, a trellis termination approach is more usually adopted, i.e. the trellis is brought to state 0 at the end of the data sequence. Termination is not as easy as for a nonrecursive code where all that is needed is a sufficient number of zeros at the end of the data. In the case of the RSC code, the flushing data depends on the state. For example, from Figure 10.12 we see that for the (1, 5/7) code the flushing sequence from state 11 is 0 1.

A single RSC encoder can easily be cleared by applying, to the input, a bit identical to the feedback bit for v cycles of the encoder (v is the memory constraint length). In this way the bits fed into the encoder memory will be zero and the shift registers will clear. However the two encoders will be in different states because of the pseudorandom interleaving of the data and therefore the effective clearing sequence will be different for the two encoders. This will not matter as long as, at the end of the encoded data, the flushing bits and generated parities for both encoders are transmitted. For our example code, each encoder would need two flushing bits and would generate two parities. Transmitting all of those would result in eight tail bits.

One interleaver design that allows simultaneous flushing of both encoders is the simile interleaver. It is based on an observation that the content of each register in the encoder is computed as the modulo-2 sum of specific series of bits, where each series consists of the bits whose position mod-$(v + 1)$ is the same. Thus for $v = 2$, bits 0, 3, 6, 9, etc. form a series as do bits 1, 4, 7, 10, etc. and bits 2, 5, 8, 11, etc. The order of the bits within each series is not important. Therefore if we can ensure that these series are separately interleaved, as shown in Figure 10.20, the end state of both encoders will be the same.

Simile interleaving can easily be achieved by a helical interleaver. This is a block interleaver with a diagonal readout order. If the data is written into columns, then the number of rows must be a multiple of $v + 1$. The number of columns and the number of rows must be relatively prime. For example, a 6×7 matrix could be used as shown in Figure 10.21.

The readout order is i_{41} i_{34} i_{27} i_{20} i_{13} i_6 i_5 i_{40} i_{33} i_{26} i_{19} i_{12} i_{11} i_4 i_{39} i_{32} i_{25} i_{18} i_{17} i_{10} i_3 i_{38} i_{31} i_{24} i_{23} i_{16} i_9 i_2 i_{37} i_{30} i_{29} i_{22} i_{15} i_8 i_1 i_{36} i_{35} i_{28} i_{21} i_{14} i_7 i_0.

Given that odd–even interleaving is just a form of simile interleaving, it is seen that if the helical interleaver has an even number of rows, as shown in this example, it has the odd–even property.

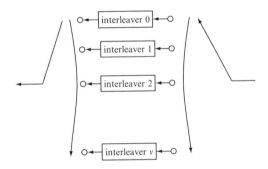

Figure 10.20   Simile interleaving

i_{41}	i_{35}	i_{29}	i_{23}	i_{17}	i_{11}	i_5
i_{40}	i_{34}	i_{28}	i_{22}	i_{16}	i_{10}	i_4
i_{39}	i_{33}	i_{27}	i_{21}	i_{15}	i_9	i_3
i_{38}	i_{32}	i_{26}	i_{20}	i_{14}	i_8	i_2
i_{37}	i_{31}	i_{25}	i_{19}	i_{13}	i_7	i_1
i_{36}	i_{30}	i_{24}	i_{18}	i_{12}	i_6	i_0

Figure 10.21   Information in 6×7 block

10.8   THE SOFT-OUTPUT VITERBI ALGORITHM

A lower complexity, but lower performance, alternative to MAP decoding is to use a soft-output version of the Viterbi algorithm [5]. Given that the Viterbi algorithm gives maximum likelihood decoding on a trellis, it may seem strange that the soft-output version leads to a lower performance. The reason is that, apart from the approximations used in deriving the soft decisions, the Viterbi algorithm minimizes the sequence error probability, whereas MAP minimizes output bit error probability. In other words, if we used both algorithms to decode the same sequences, VA would go wrong less often, but MAP would produce fewer bit errors. As a result, a MAP decoder passes fewer bit errors to the next decoder, producing more gain in each half iteration.

The SOVA decoder can be considered to start from the result of conventional hard-output Viterbi decoding and then to calculate the reliability metrics for each decoded bit. However, to implement it in that way would require retention of complete information about the paths considered during the decoding, losing some of the value of the Viterbi algorithm. Therefore it may be better to calculate the reliability metrics for each survivor path as the decoding proceeds, even though many of the calculations will be wasted as paths are subsequently eliminated.

We start with all reliabilities set to a high value. For any survivor path, we consider also the path that it superseded (called the concurrent path). We find the difference in their path metric and trace back the concurrent and survivor paths to the point where they diverged. Now we determine for which frames the two paths differ in the associated information bit and use the metric difference as the reliability of those bits unless we have already stored a lower reliability value for that bit.

Consider the example of Viterbi decoding shown in Figure 10.22. This is in fact the same as the soft-decision example given in Chapter 2 (Section 2.10), except that the mapping to information has been modified to be correct for the (1, 5/7) RSC code. Initially the reliability values are set to ∞ for each bit.

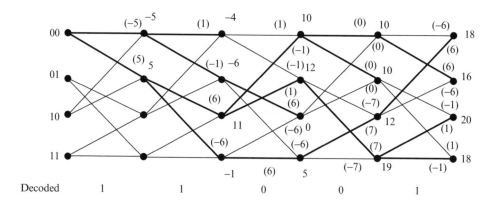

Figure 10.22   Viterbi decoding example

We construct a table as before to hold data for the Viterbi decoder, but we need an extra location for reliabilities. Because the end state cannot simply be read from the data path history, it has been shown in the tables below although it does not need to be stored in reality. After two frames, where the sequence $+3.5$ $+1.5$, $+2.5$ -3.5 is received, the paths and metrics are as shown in Table 10.1. There have been no merging paths; therefore, no reliability information can be calculated.

In the third frame we receive $+2.5$ -3.5. We now extend the paths as shown in Table 10.2.

Now we need to do the path merging and calculate the reliabilities. For the two paths merging at state 00 the difference in path metrics is 13. We choose the survivor as for the normal Viterbi algorithm but use the value 13 as the reliability for every bit where the two merged paths differ, provided that value is less than the currently stored reliability value. In this case the data differences are in frames 1, 2 and 3, giving reliability 13 for each of those bits.

Looking at the paths merging in state 01, the metric difference is 17 and the data differences are again in frames 1, 2 and 3. For the paths merging in state 10, the metric difference is 7 and the data differences are also in frames 1, 2 and 3. For the paths merging in state 11, the metric difference is 17 and the data differences are again in frames 1, 2 and 3. These leaves the SOVA table with the contents shown in Table 10.3.

Table 10.1   SOVA table after two frames

End state	Data	Path metric	Reliabilities
00	00	-4	∞ ∞
01	01	-6	∞ ∞
10	11	11	∞ ∞
11	10	-1	∞ ∞
00	00	-4	∞ ∞
01	01	-6	∞ ∞
10	11	11	∞ ∞
11	10	-1	∞ ∞

Table 10.2   SOVA table after third frame extensions

End state	Data	Path metric	Reliabilities
00	111	10	∞ ∞ ∞
01	110	12	∞ ∞ ∞
10	100	-7	∞ ∞ ∞
11	101	5	∞ ∞ ∞
00	000	-3	∞ ∞ ∞
01	001	-5	∞ ∞ ∞
10	011	0	∞ ∞ ∞
11	010	-12	∞ ∞ ∞

In the fourth frame, we receive −3.5 +3.5. We now extend the paths as shown in Table 10.4.

The metric differences to states 00 and 01 are 10 and 10, respectively, and they apply in each case to frames 1 and 4, except where the currently stored value is lower. The metric differences to states 10 and 11 are 7 and 21, respectively, and they apply in each case to frames 2, 3 and 4, except where the currently stored values are lower. Therefore after merging, the decoder table is as shown in Table 10.5.

Finally, we receive in the fifth frame, the values +2.5 +3.5 and create the path extensions shown in Table 10.6.

Table 10.3   SOVA table after frame 3 merging

End state	Data	Path metric	Reliabilities
00	111	10	13 13 13
01	110	12	17 17 17
10	011	0	7 7 7
11	101	5	17 17 17
00	111	10	13 13 13
01	110	12	17 17 17
10	011	0	7 7 7
11	101	5	17 17 17

Table 10.4   SOVA table after frame 4 extensions

End state	Data	Path metric	Reliabilities
00	0111	0	7 7 7 ∞
01	0110	0	7 7 7 ∞
10	1010	12	17 17 17 ∞
11	1011	−2	17 17 17 ∞
00	1110	10	13 13 13 ∞
01	1111	10	13 13 13 ∞
10	1101	5	17 17 17 ∞
11	1100	19	17 17 17 ∞

Table 10.5   SOVA table after frame 4 merging

End state	Data	Path metric	Reliabilities
00	1110	10	10 13 13 10
01	1111	10	10 13 13 10
10	1010	12	17 7 7 7
11	1100	19	17 17 17 21
00	1110	10	10 13 13 10
01	1111	10	10 13 13 10
10	1010	12	17 7 7 7
11	1100	19	17 17 17 21

Table 10.6   SOVA table after frame 5 extensions

End state	Data	Path metric	Reliabilities
00	10101	18	17 7 7 7 ∞
01	10100	6	17 7 7 7 ∞
10	11000	20	17 17 17 21 ∞
11	11001	18	17 17 17 21 ∞
00	11100	4	10 13 13 10 ∞
01	11101	16	10 13 13 10 ∞
10	11111	9	10 13 13 10 ∞
11	11110	11	10 13 13 10 ∞

Table 10.7   SOVA table after frame 5 merging

End state	Data	Path metric	Reliabilities
00	10101	18	17 7 7 7 14
01	11101	16	10 10 13 10 10
10	11000	20	17 17 11 11 11
11	11001	18	17 17 7 7 7
00	10101	18	17 7 7 7 14
01	11101	16	10 10 13 10 10
10	11000	20	17 17 11 11 11
11	11001	18	17 17 7 7 7

The metric differences to states 00 and 01 are 14 and 10, respectively, and apply to frames 2 and 5 unless the stored reliability is lower. The metric differences to states 10 and 11 are 11 and 7, respectively, and apply to frames 3, 4 and 5 unless the stored reliability is lower. After merging the decoder table is as shown in Table 10.7.

At the end of this example the path terminating at state 00 is selected and the output bit values are $+17$ -7 $+7$ -7 $+14$.

The performance of SOVA in iterative decoding is improved by normalization of the received bit values and of the extrinsic information passed to the next stage of decoding. This is achieved by scaling all the values so that the maximum magnitude is at some predefined level. The extrinsic information is, however, given some weighting (>1) relative to the received bits to assist with the convergence. The relative weighting is known as the *fudge factor* (because it has only an empirical validity) and the process can be thought of as normalizing the received bits to 1 and the extrinsic information to the value of the fudge factor.

10.9 GALLAGER CODES

Gallager codes [6, 7] consist of an arrangement of low density parity check (LDPC) codes suitable for iterative decoding. The LDPC codes are block codes whose parity check matrix **H** is sparse, i.e. it consists of a number of low-weight rows. The codes are usually subject to a regularity constraint, namely that each row is of fixed weight w_r and each column is of fixed weight w_c. In other words, each bit is involved in w_c parity checks and each of those checks operates over w_r bits. The proportion of bits with value 1 is ρ (a small value). Given that each column of the parity check matrix has $n - k$ bits, we can see that

$$w_c = \rho(n - k) \tag{10.2}$$

Similarly, given that each row of the parity check matrix has n bits, we can see that

$$w_r = \rho n \tag{10.3}$$

Note that, as $n - k < n$, $w_c < w_r$. From Equations (10.2) and (10.3) we find that $w_c/w_r = 1 - R$, where R is the rate of the code; hence

$$R = 1 - w_c/w_r \tag{10.4}$$

The codes are commonly denoted as (n, w_c, w_r) codes. As implied by Equation (10.4), once these three parameters are known, the number of parity checks can also be deduced. Relaxing the regularity constraint has, however, been found to lead to better codes [8].

The generation of the parity check matrix can be random but subject to a constraint that ensures a generator matrix for the code can be derived. There needs to be some matrix **H'**, consisting of $n - k$ columns of **H**, that is nonsingular, i.e. **H'** can be inverted. Note that the requirement for **H'** to be nonsingular implies that w_c is odd. An even value would mean that the modulo-2 sum of every column would be zero, implying that each row would be the modulo-2 sum of the other rows. Hence the rows would not be linearly independent.

Assume that after some rearrangement of rows, the parity check matrix is in a form

$$H^P = [P|H']$$

where H^P is a permuted form of H, and P is a $(n - k) \times n$ array of elements. Then

$$H'^{-1}H^P = \left[H'^{-1}P|I\right]$$

From this a systematic generator matrix can be derived. After encoding the bit permutation originally applied must be reversed.

Let us take as an example a case with $w_c = 3$ and $w_r = 6$ to produce a rate $1/2$ code. The 6×6 matrix

$$H' = \begin{bmatrix} 0 & 0 & 1 & 1 & 0 & 1 \\ 0 & 1 & 0 & 0 & 0 & 0 \\ 0 & 0 & 0 & 1 & 1 & 0 \\ 1 & 0 & 1 & 0 & 1 & 0 \\ 1 & 1 & 0 & 0 & 1 & 1 \\ 1 & 1 & 1 & 1 & 0 & 1 \end{bmatrix}$$

is nonsingular and has columns of weight 3. Its inverse is

$$H'^{-1} = \begin{bmatrix} 1 & 1 & 0 & 0 & 0 & 1 \\ 0 & 1 & 0 & 0 & 0 & 0 \\ 0 & 0 & 1 & 0 & 1 & 1 \\ 1 & 1 & 0 & 1 & 1 & 0 \\ 1 & 1 & 1 & 1 & 1 & 0 \\ 0 & 1 & 1 & 1 & 0 & 1 \end{bmatrix}$$

We now create a 6×12 matrix ensuring that the additional columns are also of weight 3 and that the rows have total weight 6:

$$H = \begin{bmatrix} 1 & 1 & 0 & 1 & 0 & 0 & 0 & 0 & 1 & 1 & 0 & 1 \\ 1 & 1 & 1 & 1 & 1 & 0 & 0 & 1 & 0 & 0 & 0 & 0 \\ 1 & 0 & 1 & 0 & 1 & 1 & 0 & 0 & 0 & 1 & 1 & 0 \\ 0 & 0 & 0 & 1 & 1 & 1 & 1 & 0 & 1 & 0 & 1 & 0 \\ 0 & 1 & 1 & 0 & 0 & 0 & 1 & 1 & 0 & 0 & 1 & 1 \\ 0 & 0 & 0 & 0 & 0 & 1 & 1 & 1 & 1 & 1 & 0 & 1 \end{bmatrix}$$

In systematic form the parity check matrix is

$$H_{sys} = \begin{bmatrix} 0 & 0 & 1 & 0 & 1 & 1 & 1 & 0 & 0 & 0 & 0 & 0 \\ 1 & 1 & 1 & 1 & 1 & 0 & 0 & 1 & 0 & 0 & 0 & 0 \\ 1 & 1 & 0 & 0 & 1 & 0 & 0 & 0 & 1 & 0 & 0 & 0 \\ 0 & 1 & 0 & 1 & 0 & 1 & 0 & 0 & 0 & 1 & 0 & 0 \\ 1 & 1 & 1 & 1 & 1 & 0 & 0 & 0 & 0 & 0 & 1 & 0 \\ 0 & 1 & 0 & 0 & 1 & 1 & 0 & 0 & 0 & 0 & 0 & 1 \end{bmatrix}$$

and the generator matrix is

$$G = \begin{bmatrix} 1 & 0 & 0 & 0 & 0 & 0 & 0 & 1 & 1 & 0 & 1 & 0 \\ 0 & 1 & 0 & 0 & 0 & 0 & 0 & 1 & 1 & 1 & 1 & 1 \\ 0 & 0 & 1 & 0 & 0 & 0 & 1 & 1 & 0 & 0 & 1 & 0 \\ 0 & 0 & 0 & 1 & 0 & 0 & 0 & 1 & 0 & 1 & 1 & 0 \\ 0 & 0 & 0 & 0 & 1 & 0 & 1 & 1 & 1 & 0 & 1 & 1 \\ 0 & 0 & 0 & 0 & 0 & 1 & 1 & 0 & 0 & 1 & 0 & 1 \end{bmatrix}$$

The structure of the code may be represented by a bipartite graph of the code, known as a Tanner graph [9, 10]. For our example, the graph contains twelve variable nodes on one side, six check nodes on the other side with each variable node connected to three check nodes and each check node connected to six variable nodes, as shown in Figure 10.23.

The graph shows the way in which belief about symbol values propagates through the code. Ideally the cycles (i.e. going between variable and check nodes without retracing a path and returning to the start point) in the Tanner graph should be long. The minimum cycle length is 4 (implying that two variable nodes are connected to the same two parity check nodes) and constructions aim for a higher minimum value.

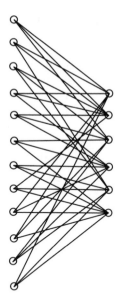

Figure 10.23 Example Tanner graph

The decoding algorithm operates alternately over rows and columns to find the most likely code vector $\mathbf{c}$ that satisfies the condition $\mathbf{cH}^T = 0$. Let $V(i)$ denote the set of w_r bits that participate in check i. Let $C(j)$ denote the set of w_c checks that check bit j. The probabilities that bit j is 0 or 1, given the parity checks other than check i, are written P_{ij}^0 and P_{ij}^1. The values are initialized to the *a priori* probabilities p_j^0 and p_j^0 of bit values 0 and 1 for each bit j (i.e. the initial values are the same for all values of i). The probabilities that check i is satisfied by a value 0 or 1 in bit j given the current values of the other bits in $V(i)$, are denoted Q_{ij}^0 and Q_{ij}^1.

The horizontal (row) step is:

Define $\Delta P_{ij} = P_{ij}^0 - P_{ij}^1$.

For each i compute ΔQ_{ij} as the product of $\Delta P_{ij'}$ for all $j' \neq j$.

Set $Q_{ij}^0 = 1/2\,(1 + \Delta Q_{ij})$, $Q_{ij}^1 = 1/2\,(1 - \Delta Q_{ij})$.

The vertical (column) step is:

For each j, compute P_{ij}^0 as p_j^0 times the product of $Q_{i'j}^0$ for all $i' \neq i$ and P_{ij}^1 as p_j^1 times the product of $Q_{i'j}^1$ for all $i' \neq i$. Scale the values of P_{ij}^0 and P_{ij}^1 such that $P_{ij}^0 + P_{ij}^1 = 1$.

For each j, compute P_j^0 as p_j^0 times the product of Q_{ij}^0 for all i and P_j^1 as p_j^1 times the product of Q_{ij}^1 for all i. Scale the values of P_j^0 and P_j^1 such that $P_j^0 + P_j^1 = 1$.

The values of P_j^0 and P_j^1 are hard-decision quantized and used to check the condition $\mathbf{cH}^T = 0$. If this fails then the values of P_{ij}^0 and P_{ij}^1 are fed back to another iteration of the horizontal step. This continues for as many iterations as desired.

Decoding example

Suppose our received sequence has values of p^1 equal to 0.7 0.1 0.4 0.1 0.1 0.6 0.9 0.2 0.8 0.1 0.1 0.1. The hard-decision quantized form of this is 100001101000 and it fails the syndrome test. We therefore undertake decoding. The values of P_{ij}^1 and P_{ij}^0 are shown in Tables 10.8 and 10.9.

We now define the values of ΔP_{ij} as shown in Table 10.10.

The row decoding calculations of ΔQ_{ij} give the values shown in Table 10.11 from which the values of Q_{ij}^0 and Q_{ij}^1 are as shown in Tables 10.12 and 10.13, respectively.

Table 10.8 Values of P_{ij}^1

0.7	0.1		0.1					0.8	0.1		0.1	
0.7	0.1	0.4	0.1	0.1			0.2					
0.7		0.4		0.1	0.6				0.1	0.1		
			0.1	0.1	0.6	0.9		0.8		0.1		
	0.1	0.4				0.9	0.2			0.1	0.1	
					0.6	0.9	0.2	0.8	0.1		0.1	

Table 10.9 Values of P_{ij}^0

0.3	0.9		0.9					0.2	0.9		0.9	
0.3	0.9	0.6	0.9	0.9			0.8					
0.3		0.6		0.9	0.4				0.9	0.9		
			0.9	0.9	0.4	0.1		0.2		0.9		
	0.9	0.6				0.1	0.8			0.9	0.9	
					0.4	0.1	0.8	0.2	0.9		0.9	

Table 10.10 Values of ΔP_{ij}

−0.4	0.8		0.8					−0.6	0.8		0.8	
−0.4	0.8	0.2	0.8	0.8			0.6					
−0.4		0.2		0.8	−0.2				0	0.8	0.8	
			0.8	0.8	−0.2	−0.8	0	−0.6		0.8		
	0.8	0.2				−0.8	0.6			0.8	0.8	
					−0.2	−0.8	0.6	−0.6	0.8		0.8	

Table 10.11 Values of ΔQ_{ij}

−0.25	0.12		0.12					−0.16	0.12	0.00	0.12	
0.06	−0.03	−0.12	−0.03	−0.03			−0.04					
−0.02		0.04		0.01	−0.04					0.01	0.01	
			−0.06	−0.06	0.25	0.06		0.08		−0.06		
	−0.06	−0.25				0.06	−0.08			−0.06	−0.06	
					0.18	0.05	−0.06	0.06	−0.05		−0.05	

We now carry out column decoding. The calculations of P_{ij}^0 and P_{ij}^1 (before scaling) give the values as shown in Tables 10.14 and 10.15, respectively.

To check whether we need to carry out another iteration, we calculate P_j^0 and P_j^1 and, after scaling, obtain the values as shown in Tables 10.16 and 10.17, respectively.

The hard-decision quantized sequence is 1 0 1 0 0 0 1 0 1 0 0 0. This satisfies the syndrome test, indicating that decoding is complete.

Table 10.12 Values of Q_{ij}^0

0.38	0.56		0.56					0.42	0.56		0.56
0.53	0.48	0.44	0.48	0.48				0.48			
0.49		0.52		0.51	0.48				0.51	0.51	
			0.47	0.47	0.62	0.53		0.54			0.47
	0.47	0.38				0.53	0.46			0.47	0.47
					0.59	0.52	0.47	0.53	0.48		0.48

Table 10.13 Values of Q_{ij}^1

0.62	0.44		0.44					0.58	0.44		0.44
0.47	0.52	0.56	0.52	0.52				0.52			
0.51		0.48		0.49	0.52				0.49	0.49	
			0.53	0.53	0.38	0.47		0.46			0.53
	0.53	0.62				0.47	0.54			0.53	0.53
					0.41	0.48	0.53	0.47	0.52		0.52

Table 10.14 Values of P_{ij}^0 from column decoding

0.08	0.20		0.20					0.06	0.22		0.20
0.06	0.24	0.12	0.24	0.21				0.17			
0.06		0.10		0.20	0.15				0.24	0.20	
			0.24	0.22	0.11	0.03		0.04			0.21
	0.24	0.14				0.03	0.18			0.21	0.24
					0.12	0.03	0.18	0.05	0.26		0.24

Table 10.15 Values of P_{ij}^1 from column decoding

0.17	0.03		0.03					0.17	0.03		0.03
0.22	0.02	0.12	0.02	0.03				0.06			
0.20		0.14		0.03	0.09				0.02	0.03	
			0.02	0.03	0.13	0.20		0.22			0.03
	0.02	0.11				0.20	0.06			0.03	0.02
					0.12	0.20	0.06	0.21	0.02		0.02

Table 10.16 Values of P_j^0 from column decoding

0.22	0.91	0.44	0.91	0.88	0.60	0.13	0.73	0.19	0.91	0.88	0.90

Table 10.17   Values of P_j^1 from column decoding

0.78	0.09	0.56	0.09	0.12	0.40	0.87	0.27	0.81	0.09	0.12	0.10

10.10   SERIAL CONCATENATION WITH ITERATIVE DECODING

Serially concatenated codes are suitable for iterative decoding [11, 12]. For example, RSC codes can be serially concatenated as shown in Figure 10.24. The transmitted stream contains:

1   The information plus tail bits to clear encoder 1.

2   The interleaved parity bits from encoder 1 plus the tail bits to clear encoder.

3   The parity bits from encoder 2.

At the decoder, the procedure is to decode the second code and deinterleave to provide parity bits for decoder 1. Decoder 1 produces not only the decoded information but also a new estimate of the parity bits from encoder 1. These are then interleaved to feed back to decoder 2 for the next iteration. This is shown in Figure 10.25. For simplicity, the derivation of extrinsic information from each half iteration has not been shown.

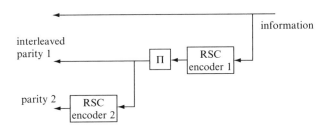

Figure 10.24   Serial concatenation of RSC codes

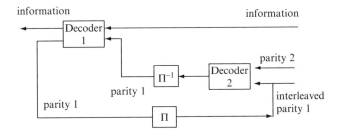

Figure 10.25   Iterative decoding of serially concatenated RSC codes

10.11   PERFORMANCE AND COMPLEXITY ISSUES

Iterative decoding requires the component codes to be simple because of the repetitions of the decoding operations. In reality the algorithms, particularly BCJR, are complex, although the number of states in the RSC trellis is usually small (no more than 16). The performance gains are substantial, however, and performance within 0.5 dB of the Shannon limit is possible. This means that with rate 1/2 encoded QPSK, we can get low post-decoding bit error rates (around 10^{-5}) for $E_b/N_0 < 0.5$ dB.

There are several qualifiers that need to be entered to the above statement. The first is that the block length affects performance because the best interleaving can be achieved with long blocks. The second is that the number of iterations may be large, although there are decreasing returns beyond about 10 iterations. Of course each half iteration introduces delay approximately equal to one block interval, so constraints on latency will limit the block length and number of iterations. It is easy to lose 1 dB or more of performance through such limitations. Gallager codes are simpler on each iteration, but the number of iterations used is higher (100 or more) and the performance is not quite as good.

The actual performance curves follow a general form shown in Figure 10.26. Above a certain value of E_b/N_0 the output bit error rates fall very rapidly – the so-called waterfall or convergence region. However the codes contain a small number of low-weight sequences that limit the asymptotic performance (performance at high signal-to-noise ratios). The result is a flattening of the BER curve creating an error floor often at around 10^{-6}. This is largely an interleaver problem and several methods have been proposed to overcome this limitation, some of which are mentioned in the final section of this chapter.

To understand how interleaver design affects the existence of low-weight codewords, consider the example code of Section 10.5 with the state diagram given in Figure 10.12. A sequence which is all-zero except for a section 1001 at the input of the encoder will produce a nonzero portion of the code sequence 11 01 01 11, which is weight 6 and returns the encoder to state 00. If this data is put into a regular block interleaver, as described in Chapter 1 (Section 1.12), the interleaved encoder will see a long interval between the 1s during which time it will output a long nonzero sequence. Thus this low weight input will result in a high-weight code sequence. However with the same interleaver it is easy to arrange a weight 4 input sequence so that two rows and two columns of the interleaver each contain the sequence 1001.

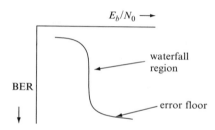

Figure 10.26   General form of iterative decoding performance curves

As a result each encoder will produce two nonzero portions of weight 6, giving a weight 24 code sequence. To eliminate this weight 24 sequence will require some irregularity in the interleaver pattern.

For the best performance from the BCJR algorithm, the signal strength and noise levels need to be known for accurate setting of the log-likelihood ratios. Channel estimation methods are therefore needed and the estimates themselves will inevitably contain inaccuracies. Fortunately it is found that the effect of overestimating signal-to-noise ratio is much less serious than underestimating, so estimators can be biased to overcome the effects of inaccuracy.

10.12   APPLICATION TO MOBILE COMMUNICATIONS

The third generation mobile communication system UMTS (Universal Mobile Telephone System) specifies for its radio network UTRAN (UMTS Radio Access Network) a parallel concatenated 8-state RSC code [13], shown in Figure 10.27. Pseudorandom interleaving is used within the blocks of data; their length can vary from 40 to 5114 bits if frequency division duplexing of uplink and downlink is used. No puncturing is used, resulting in a rate 1/3 code. At the end of the data both encoders are cleared and the clearing sequences are transmitted along with the parity bits generated. This means that there is a 12-bit tail sequence on each transmission.

Simulated performance of these codes for 1600 bits of data and four iterations is shown in Figure 10.28 [14]. The curves show the results of both *log*-MAP and SOVA decoding for AWGN and uncorrelated Rayleigh fading channels.

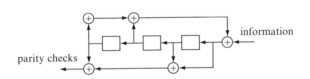

Figure 10.27   UMTS RSC encoder

10.13   TURBO TRELLIS-CODED MODULATION

There are several ways in which turbo codes can be applied to MPSK or QAM modulations to produce schemes that are both bandwidth efficient and have large coding gains. Possible approaches are to use standard binary turbo codes mapped onto the multilevel modulation [15], to use a number of parallel codes (in this case turbo codes) mapped onto the modulation [16] or to use parallel concatenation of Ungerboeck codes converted to RSC form [17–19].

The first approach is known as the pragmatic approach. For example, a rate 1/3 turbo code could be punctured to rate 2/3 and applied, using Gray code mapping, to an 8-PSK constellation. Another example for 16-QAM would be to puncture to rate 1/2

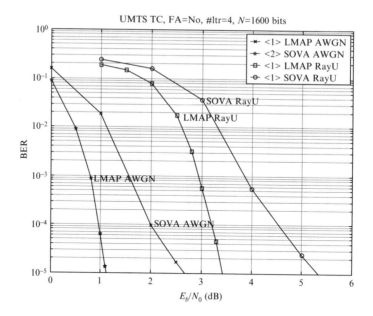

Figure 10.28 Performance of UMTS turbo codes

and then map one information bit and one parity bit to each of the I and Q channels (also Gray coded). An interleaver could be used prior to the mapping onto the modulation to combat burst errors on the channel. There are some nonoptimum approximations used in the assumption of independence when the log-likelihood ratios of the different bits are calculated at the receiver; nevertheless such schemes can be made to work well and the same turbo codes can be applied, with different mappings, to different modulations.

For the second approach, we provide one turbo encoder for each bit of the multilevel modulation, e.g. 3 for 8-PSK or 4 for 16-QAM. The data is demultiplexed into the encoders and the encoded outputs fed into a signal mapper. Decoding uses a multistage approach in which one code is decoded and the decisions used to assist the decisions of all subsequent decoders. With the right choice of component codes and code rates, this approach also works well [20].

For the third approach, we take an Ungerboeck convolutional code and convert it into RSC form. For example, the encoder of Figure 10.29 is equivalent to the rate 2/3 8-state code for 8-PSK described in Chapter 2, Section 2.15. The information is

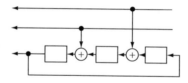

Figure 10.29 RSC equivalent of rate 2/3 Ungerboeck code for 8-PSK

encoded directly and, after interleaving, by a second RSC encoder (using the same code). The overall code rate is now half that of the Ungerboeck code, a problem we solve by puncturing. The output of the second encoder could be deinterleaved so that the information bits are the same as for the first encoder and then the choice of parity bits is alternated between the two encoders. Alternatively if the interleaver has an odd–even property we could choose alternately between the entire output frames of the two encoders knowing that all information bits will appear in the transmitted stream.

10.14   CONCLUSION

Iteratively decoded codes are the most recent and most exciting development in coding. At the time of writing the subject has found its way into three good texts on digital communications [21–23], one general purpose coding text [24] and two specialist texts on turbo codes [25–26].

   In such a rapidly developing field, new ideas are being advanced all the time to improve performance, particularly to investigate and lower the error floor. Approaches to this include special design of interleavers for specific codes, irregular interleavers [27] and woven convolutional codes designed to have good distance properties [28]. Nonbinary RSC codes [29] appear to be attractive to produce good high rate coding schemes. Iteratively decoded product codes, known as turbo product codes (TPC), are also of considerable interest [30–32].

10.15   EXERCISES

1   Find the MAP decoding of the sequence 1.2, 0.3, 0.8, −1.1, 0.7, −1.2, −1.0, 1.1, 0.9, 1.1 for the example RSC code of Section 10.5.

2   Find the SOVA decoding of the sequence +3.5, +2.5, +0.5, −2.5, −0.5, −2.5, −3.5, +3.5, −1.5, +3.5, +2.5, −0.5 for the example RSC code of Section 10.5.

3   A rate 1/3 turbo code has lowest weight nonzero code sequences of weight 24. It is applied to a QPSK modulation. Estimate the BER at which the error floor will become apparent.

4   Considering a block interleaved turbo code as in Section 10.11, find the weight of the sequence produced by data 111 input to the example RSC encoder of Section 10.5. Hence find the weight of the code sequences resulting from a weight 9 data pattern with 111 on each of three rows and three columns. Hence deduce whether input or output weight at a single RSC encoder is more important in identifying low-weight turbo code sequences.

5   Find the lowest weight sequences in a turbo code using regular block interleaving and component $(1, 7/5)$ RSC codes. Comment on the comparison with the $(1, 5/7)$ RSC code and on the desirable characteristics of the feedback polynomial.

10.16 REFERENCES

1 L.R. Bahl, J. Cocke, F. Jelinek and J. Raviv, 'Optimal decoding of linear codes for minimizing symbol error rate', IEEE Transactions on Information Theory, Vol. IT-20, pp. 284–287, 1974.

2 C. Berrou, A. Glavieux and P. Thitmajshima, *Near Shannon limit error-correction coding and decoding: Turbo codes*, Proceedings of 1993 International Conference on Communications, pp. 1064–1090, Geneva, Switzerland, May 1993.

3 C. Berrou and A. Glavieux, *Near Shannon limit error-correction coding and decoding: Turbo codes*, IEEE Transactions on Communications, Vol. 44, pp. 1261–1271, 1996.

4 P. Robertson, E. Villebrun and P. Hoeher, *A comparison of optimal and sub-optimal MAP decoding algorithms operating in the log domain*, Proc. ICC'95, Seattle, June 1995.

5 J. Hagenauer and P. Hoeher, 'A Viterbi algorithm with soft-decision outputs and its application,' in Conf. Proceedings IEEE Globecom '89, pp. 1680–1686, 1989.

6 R.G. Gallager, *Low Density Parity Check Codes*, MIT Press, 1963.

7 D.J.C. MacKay, *Good error correcting codes based on very sparse matrices*, IEEE Transactions on Information Theory, Vol. 45, No. 2, pp. 399–431, March 1999.

8 T. Richardson, A. Shokrollahi and R. Urbanke, *Design of capacity approaching irregular low-density parity check codes*, IEEE Transactions on Information Theory, Vol. 47, No. 2, pp. 619–637, February 2001.

9 R.M. Tanner, *A recursive approach to low-complexity codes*, IEEE Transactions on Information Theory, Vol. 27, pp. 533–547, 1981.

10 N. Wiberg, H.-A. Loeliger, and R. Kötter, *Codes and iterative decoding on general graphs*, Euro. Trans. Telecommun., Vol. 6, pp. 513–526, 1995.

11 A. Ambroze, G.Wade, M. Tomlinson, *Iterative MAP decoding for serial concatenated convolutional codes*, IEE Proceedings Communications, Vol. 145, No. 2, pp. 53–59, April 1998.

12 K.R. Narayanan, G.L. Stuber, *A Serial Concatenation Approach to Iterative Demodulation and Decoding*, IEEE Transactions on Communications, Vol. 47, No. 7, pp. 956–960, July 1999.

13 3GPP TS25.212 v3.4.0 Multiplexing and Channel Coding (FDD).

14 Y. Rosmansyah, *Channel and decoding effects on UMTS turbo code performance*, CCSR internal report, University of Surrey, U.K., August 2001.

15 S. LeGoff, A. Glavieux and C. Berrou, *Turbo codes and high efficiency modulation*, Proceedings of IEEE ICC'94, pp. 645–649, New Orleans, May 1994.

16 H. Imai and S. Hirakawa, *A new multilevel coding method using error correcting codes*, IEEE Transactions on Information Theory, Vol. 23, No. 3, pp. 371–377, May 1977.

17 P. Robertson and T. Wörtz, *Novel coded modulation scheme employing turbo codes*, IEE Electronics Letters, Vol. 31, No. 18, pp. 1546–1547, August 1995.

18 P. Robertson and T. Wörtz, *Bandwidth efficient turbo trellis-coded modulation using punctured component codes*, IEEE Journal on Selected Areas in Communications, Vol. SAC-16, No. 2, pp. 206–218, February 1999.

19 S. Benedetto, D. Divsalar, G. Montorsi and F. Pollara, *Parallel concatenated trellis coded modulation*, Proceedings IEEE ICC'96, pp. 974–978, 1996.

20 L.U. Wachsmann and J. Huber, *Power and bandwidth efficient digital communication using turbo codes in multilevel codes*, European Transactions in Telecommunications, Vol. 6, No. 5, pp 557–567, Sept/Oct 1995.

21 J.G. Proakis, *Digital Communications* McGraw Hill, 2001.

22 S. Haykin, *Communication Systems*, John Wiley & Sons, 2001.

23 B. Sklar, *Digital Communications – Fundamentals and Applications*, Prentice Hall, 2001.

24 M. Bossert, *Channel Coding for Telecommunications*, John Wiley & Sons, 1999.

25 C. Heegard and S.B. Wicker, *Turbo Coding*, Kluwer Academic Publishers, 1998.

26 B. Vucetic and J. Yuan, 'Turbo Codes – Principles and Applications', Kluwer Academic Publishers, 2000.

27 B.J. Frey and D.J.C. MacKay, *Irregular turbocodes*, Proceedings of the 37th Annual Allerton Conference on Communications, Control and Computing, Allerton House, Illinois, September 1999.

28 J. Freudenberger, M. Bossert, S. Shavgulidze, and V. Zyablov, *Woven codes with outer warp: variations, design, and distance properties*, IEEE Journal on Selected Areas in Communications, pp. 813–824, May 2001.

29 C. Berrou and M. Jézéquel, *Nonbinary convolutional codes for turbo coding*, IEE Electronics Letters, Vol. 35, No. 1, pp. 39–40, January 1999.

30 R.M. Pyndiah, *Near optimum decoding of product block codes: block turbo codes*, IEEE Trans Commun 1998, Vol. 46, No. 8, pp. 1003–1010.

31 Aitsab and R.M. Pyndiah, *Performance of Reed Solomon block turbo codes*, Proceedings of IEEE Globecom conference 1996, Vol. 1/3, pp. 121–125.

32 P. Sweeney and S. Wesemeyer, *Iterative Soft Decision Decoding of Linear Block Codes*, IEE Proceedings on Communications, Vol. 147, No. 3, pp. 133–136, June 2000.

Index